ON PANDEMICS

ON PANDEMICS

Deadly Diseases from Bubonic Plague to Coronavirus

DAVID WALTNER-TOEWS

GREYSTONE BOOKS
Vancouver/Berkeley

20 21 22 23 24 5 4 3 2 1

The quote on page 45 from *The Plague*, by Albert Camus, is from a translation by Stuart Gilbert, published by The Modern Library, Random House, 1948.

Greystone Books Ltd.
greystonebooks.com

Cataloguing data available from Library and Archives Canada
ISBN 978-1-77164-811-0 (pbk.)
ISBN 978-1-77164-812-7 (epub)

Editing by Nancy Flight
Copy editing by Wendy Fitzgibbons
Editing for second edition by Paula Ayer
Proofreading by Alison Strobel
Cover design by Nayeli Jimenez
Text design by Nayeli Jimenez and Naomi MacDougall
Cover photograph by Bernhard Lang/Getty Images

Printed and bound in Canada on ancient-forest-friendly paper by Friesens

Greystone Books gratefully acknowledges the Musqueam, Squamish, and Tsleil-Waututh peoples on whose land our office is located.

Greystone Books thanks the Canada Council for the Arts, the British Columbia Arts Council, the Province of British Columbia through the Book Publishing Tax Credit, and the Government of Canada for supporting our publishing activities.

Canada

For two great teachers of veterinary epidemiology—

the late John Iversen,

who inspired me to care about

the dead skunk in the middle of the road,

and Wayne Martin,

who taught me to measure its significance

From ghoulies and ghosties
And long-leggedy beasties
And things that go bump in the night,
Good Lord, deliver us!

A traditional prayer, of unknown provenance

CONTENTS

(1)

AN INTRODUCTION TO THE SECOND EDITION

I WROTE THE first edition of this book in the turbulent wake of avian influenza's spectacular appearance in the late 1990s, and the even more dramatic SARS alarm early in this century. In quite different ways, both diseases awakened fears and memories of the 1918 influenza pandemic, and led to widespread animal slaughtering, airport closings, government panic, mayoral rants, and rock concerts. On the other hand, the pandemic of influenza that swept around the world in 2009 and 2010, first called swine flu, and later given the technical moniker H1N1, evoked measured responses accompanied by widespread skepticism.

In 2020, I am writing this second edition of the book during a global lockdown in response to the explosive spread of SARS-COV-2 (the virus) and COVID-19 (the disease it causes). COVID-19 has now joined the list of pandemics, near-pandemics, and possible pandemics that have long haunted human history. With popular names such as SARS, bird flu, swine flu, Ebola, and bubonic plague, these diseases all have one thing in common: they are zoonoses, diseases that have made the jump from their natural homes in other animals to take up residence in people. Some of them—the influenzas, mostly—took the direct route from chickens or pigs to people. Others, such as Ebola, COVID-19, and SARS, took a more circuitous route from bats, pausing at rest stops in one or two other animals—civets, monkeys, or pangolins, perhaps—before they found their way to humans. Welcome to the twenty-first century.

The eruption and worldwide spread of SARS-COV-2 was, some might assert, predictable, but predictable only in the sense that earthquakes and volcanic eruptions in certain parts of the world are predictable. The Ring of Fire, for instance, home to more than three-quarters of the world's volcanoes, is a horseshoe-shaped area rimming the Pacific Ocean. We can predict that there will be volcanic eruptions and earthquakes along this rim, but exactly where and when they will occur is uncertain.

There were reports, of course, briefings, warnings about emerging infectious diseases, rumors of chimeric viruses that bore traces of their animal origins. In retrospect the reports should have been heeded; they weren't just angry jeremiads. They were based on scientific studies and simulations. But there were events other than rogue viruses in a Chinese market that kept us distracted. And in our collective distraction and denial, we were surprised. I could say we shouldn't have been surprised, but I've never found retrospective told-you-sos to be comforting or useful.

Even the most scientifically literate among us might be forgiven our skepticism that an outbreak led by an unstable, albeit crowned, virus, was going to usher in a New Age, or even a Different Age. Some might argue, based on good evidence, that crowned heads have always been somewhat unstable. On the other hand, we wouldn't be the first to have underestimated the impact of a small event—say, a butterfly flapping its wings in Brazil on storms in Texas. The world is more chaotic, less predictable, than we would like to believe.

Many of the pandemics imagined in movies and novels before 2020 involved apocalyptic scenarios that included people stumbling through garbage-strewn streets, bleeding from all orifices, zombies, billions dead, and bodies in the street. Perhaps a few religious or ideologically obsessed zealots even wished, mostly in secret, for one of those God-save-us pandemics that historian Walter Scheidel announced might provide a sufficiently violent shock to "upend the established order" and "flatten disparities in income and wealth."

What most literary scenarios did not imagine was an infection that spread rapidly, infected hundreds of thousands, and seemed to kill almost at random. Of course, as with previous pandemics, COVID-19 has tended to kill more older people, and people whose immune systems were already under duress from cancer or diabetes or heart disease. But what has struck many observers, myself included, was how two healthy people in the prime of life could be infected with SARS-COV-2, and one of them, without any reasonable medical explanation, would die, while the other lived. This almost seemed like the random drive-by shootings described by my Colombian colleagues from the 1990s.

So, yes, the SARS-COV-2 pandemic was predictable, but as one of my Italian colleagues, in lockdown as I write this, said, "Even in the case of an earthquake, whilst the earth is shaking, the first reaction (seconds or less) is denial: it is impossible, it cannot be true, not here! And even afterward, when you see the ruins, it's still hard to believe."

Most of us have been made aware of the ravaged landscapes, lost habitats, and disappearing megafauna that have characterized the past century. We fret about the extinctions of birds and rhinos. We try to save some arthropods, like bees and butterflies, even as we try to kill others, like mosquitoes. Still, even the most conservation-minded among us rarely think about the trillions of viruses, yeasts, fungi, and bacteria for whom these disappearing animals are home, or about where those microfauna might look for new homes once we raze their habitats for mines, cattle ranches, or cities. The diseases described in this book are, in a sense, about microscopic refugees from those lost habitats and disappearing species.

The living things on this planet are one big, dysfunctional extended family of species, in which most bacteria, viruses, and parasites are beneficial and necessary, in which diseases have a useful role in nature, and in which we ourselves have evolved from microbes and are composed of them. I think we—this big family, including the bothersome and wonderful and contradictory human species—can work out our problems with the help of some serious narrative therapy. War, with all its calls to arms, technological weapons, and national pride, its suspension of civil liberties, xenophobia, and collateral damage, is often used as a metaphor for how we should fight disease. But that is too impoverished and small-minded an image. Perhaps politics, the so-called art of the possible, is a better metaphor. War is a last resort.

After millennia of war on the agents of disease and the eradication of some of the worst, we might even find ways of negotiating with them, of accommodating each others' needs, engaging in minor, stylized skirmishes, with a reasonable, acceptable death toll on all sides. In the twenty-first century, we are discovering that we will have a common future, or none at all. But such a future will require us to educate ourselves in different, more ecologically aware, ways. One of our many tasks is to translate that education into a new version of common sense, a kind of solidarity with other people, and other species, informed by careful attention to this amazing planet we share.

Clinical neurologist Oliver Sacks has argued that we "do not honor our peripheral vision as much as we should." Sacks was talking about his own individual experience, but some of us have argued that the highly focused-but-disoriented stumbling about of biomedical sciences reflects a collective lack of honoring—indeed, a pathological loss—of peripheral vision.

If epidemiologists of human diseases were more aware of animal diseases, if veterinary epidemiologists spent more time having conversations with public health officials, if economists and politicians were more aware of complex social-ecological webs, if every upstart business guru were more aware of the unintended consequences of disruptive entrepreneurial innovations, if we were as good at paying attention to the world around us as we are to what is just in front of us... then maybe we would not have been so shocked by the appearance of COVID-19. Those are a lot of ifs, none of which can be addressed by any single profession or scientist or politician; peripheral vision in the global sense requires us to watch out for each other. In an overcrowded twenty-first-century world, we are each other's peripheral vision.

We will all die. That is a normal part of life. Still, there are ways to make our dying more pleasant, less catastrophic, more as if we are among mortal friends. As the poet W.H. Auden once wrote: The slogan of Hell is to eat or be eaten; the slogan of heaven is to eat and be eaten.

If we understand just a little more about the animals we share this planet with and the micro-organisms that they carry around, both those that we eat, and those that will eat us, we might even begin to understand ourselves.

FINALLY, BEFORE WE step into the butterfly-induced tornado of terminology related to outbreaks, epidemics, and pandemics, I should say something about names. If the names of hurricanes are open to criticism, the names of diseases are even more fraught. The task of naming things, whether in upper-class European Latin from the seventeenth century, or in popular culture, is a very human burden. This task is not to be undertaken without careful thought. In some cases, careless naming has led to reckless blowing up of bat caves and slaughtering of civets or dogs. Naming a disease after where it was first found (West

Nile, Lagos, Hong Kong, Asian, Russian, Wuhan) might provide useful shorthand for field investigators—or an opportunity for demagogic, xenophobic, and racist rants. A name can hinder the implementation of good public health programs. H1N1 was called swine flu, until several leaders in the Middle East, understandably offended, suggested that it be called the Mexican virus. Virologists settled the matter with a name that sounds like a postal code for an address in another, parallel, world. At other times, after someone, whether thoughtlessly or maliciously, labeled a disease according to ethnicity, sexual orientation (e.g., HIV as a "gay disease"), nationality, or economic status, millions of people have been stigmatized, ostracized, assaulted in the streets, and murdered. In this book I shall, insofar as is possible, use technical scientific names and, where these are not available or are too obtuse, descriptive popular names, such as SARS (severe acute respiratory syndrome).

(2)

THREATS, OUTBREAKS, EPIDEMICS, AND PANDEMICS: A PRIMER

UNTIL A few years ago, many scientists had banished words like "threat" and "danger" from our vocabulary. In an attempt to be more rigorously quantitative and less emotional, we began to write about risks. Our response to danger was called "risk management." A risk is a threat or a danger that you put into a box. Then you can count boxes, and manage them. The assumption in risk management is that you can quantify danger. This is only partially true.

Epidemiologists sometimes define an epidemic as more cases than they would expect, or as "unusually high rates." But measuring this isn't as simple as it sounds. We can ask, first of all, whether in our case numbers we are counting people exposed to a virus (and who may be test-positive), people who have been infected but don't show any clinical disease, people who are clinically ill, or people who die. We might, if we are the World Bank and are worried about economic impacts, try to calculate disability-adjusted life years; that is, how many years of economically productive life are lost. This means that the younger you are when you get sick, the greater the impact (assuming you are economically productive).

Once we decide who and what to count, we are faced with another set of questions. What are we referring to when we say "expected" or "usual"? That is relatively easy, at least from a scientific point of view. The expected or usual numbers are those we've seen over the past few decades. Even if the numbers of cases are unexpected or unusual, however, we need to ask: Are we dealing with a disease that is important?

How do we assess importance? Is it a scientific decision? Are authorities more hesitant to use the loaded word "pandemic" for some diseases, preferring to speak of a "global epidemic" of AIDS or "high rates" of malaria or diarrhea in certain parts of the world? If so, why is that? Do some diseases so radically and explicitly expose global economic inequities that the wealthy owners of global institutions would prefer to focus on those that more directly threaten Europe and North America and for which a technical, money-making fix is more likely to be found? Hey, I'm a curious epidemiologist. I'm just asking. The truth is that although words like "outbreak," "epidemic," and "pandemic" have a scientific ring to them, and some grounding in science, their use is very political.

Now, after decades of quantitative risk management, in the new age of emerging infectious diseases such as bird flu, SARS, Ebola virus disease, and COVID-19, we seem to be back in a jungle of threats and dangers. If SARS-COV-2 has taught us anything at all, it's that even our best quantitative, scientific measurements cannot give us all the answers we want.

Before we dive into what we mean by a pandemic—that equivalent to the man-eating lion in the dark forest—let me begin the definition with something smaller, at least in terms of numbers: an outbreak. An outbreak occurs when a relatively small group of people or animals or plants gets sick, as when everyone gets sick after eating a warm potato salad on a sunny day. The source of an outbreak can usually be traced to one particular event or exposure.

The next step up from an outbreak is an epidemic, which is like an outbreak, only huger. The word "epidemic" goes at least as far back as Homer, in about the eighth century BC. Homer used it to refer to someone in his or her own country, as differentiated from a traveler. It had connotations of "indigenous" or "endemic." Hippocrates, in 430 BC, gave it a medical slant, referring to physical syndromes (illnesses) that occurred in particular places and times. After the discovery of bacteria in the nineteenth century, people began to use the term to refer to specific diseases, as in epidemics of cholera. More recently, the word has been used to refer to more cases than expected, both of very specific diseases, such as hemolytic uremic syndrome caused by *E. coli* O157:H7, and general syndromes, such as obesity.

This brings us back to the question of expectations. Every year, we expect a certain number of cases of say, influenza A. When we get more than we expect, or we are faced with a new variation of the same old thing, we call it an epidemic. Some diseases, such as West Nile virus disease and Lyme disease, start as epidemics. They surprise us. But after a few years, we get used to them, and we think of them as troublesome, tiresome, endemic. They now belong here, wherever that "here" may be. The words we use to describe these disease patterns are both descriptive and a way to catch—or divert—our attention. Why, some might ask, are there more cases than we expect? How and why have our expectations changed?

Multi-country outbreaks of various strains of food-borne *Salmonella*, noroviruses, and *E. coli* are regularly reported around the world. These are rarely called epidemics. And almost never called pandemics. These are the "normal" costs of producing lots of food, or so we are led to believe. If nothing else, the language we use should tell us something about our expectations in the twenty-first century.

Pandemics are a step beyond epidemics, at least in terms of numbers. The World Health Organization (WHO) defines a pandemic as "an epidemic occurring worldwide, or over a very wide area, crossing international boundaries and usually affecting a large number of people." The classical definition includes nothing about severity: a disease may become pandemic without being a serious killer. A serious killer disease, even one appearing in many different parts of the world, may not necessarily be classified as a pandemic.

SARS, for instance, was never officially declared a pandemic. Our global alarm was that it *might become* a pandemic, and was intensified by a general confusion about the non-human origins of most human diseases. Similarly, HIV/AIDS spread around the world in ways that some of us could only describe as a pandemic. In fact, although many agencies and researchers do call it a pandemic, WHO refers to it as a "global epidemic." Before 2020, the only WHO-declared pandemics since 1900 were in 1918, 1958, 1968, and 2009. All were influenza, and all our guidelines for pandemic response are based on influenza.

If you find this confusing, you are not alone.

Influenza viruses afflicting millions of people annually cross international boundaries in temperate southern and northern hemispheres. However, these "seasonal epidemics" are not called pandemics. An influenza pandemic is a global outbreak of a *new* influenza A virus, such as the H1N1 strain associated with the 2009–10 pandemic. According to the Centers for Disease Control and Prevention (CDC), "Pandemics happen when *new* (novel) influenza A viruses emerge which are able to infect people easily and spread from person to person in an efficient and sustained way." This is why, when pandemics are first described, we often see the word "novel" or "variant" or just a little "n" or "v" in front of the name.

In 2009, WHO published a document called "Pandemic Influenza Preparedness and Response," which offers more specific guidance. This guidance document described a model with six pandemic phases. The model starts with Phase One, in which other animals, but no people, are infected, and finally arrives at Phase Six, the Pandemic Phase, in which there is "increased and sustained [human-to-human] transmission in the general population." To be considered a true pandemic, however, this human-to-human transmission can't just be inside one country, or even in two countries in the same WHO administrative region, but must also be in at least one country in a different administrative region.

After that, waves of the disease go around the world, but fewer people get sick and die with each subsequent wave, either because humans have built up immunity or because the agent, defying creationists, evolves through the processes of mutation and natural selection. In this case, only those with milder forms of the disease survive long enough to pass it on to others, and hence the agent moves in with our species to a longer, gentler, more sustainable life.

In 2017, WHO published another guide, called "Pandemic Influenza Risk Management." This guide is not a planning document. It is intended to "manage disaster risks" and to help countries assess risks so that they can make more informed decisions. This document, which appears to be the one that WHO used during the SARS-COV-2 pandemic, only shows four phases: Interpandemic, Alert, Pandemic, Transition and then—wait for it, another Interpandemic phase. Note that this document assumes that whenever we are not in a pandemic, we are between pandemics, as we are between ice ages. Although they were designed to manage influenza in humans, WHO phases can be applied to all infectious diseases. There is no non-pandemic phase in our future. We have always lived between pandemics and we always will.

I will remind you (and me), here, that the word "pandemic" does not imply severity. The emergence of H1N1 in pig populations and its spillover into humans appeared, at least at first, to be an example of this. In 2009 this new influenza virus was reported from Mexico and spread very rapidly into an officially declared global pandemic in just a few months. For both scientists and just plain folks, the whole thing was befuddling. Was this serious? Was it real? First bird flu. Now swine flu. Really?

To the relief and puzzlement of many, the new virus seemed to result in a human disease no more, or less, serious than the "normal flu." Admittedly, "normal" influenza infects millions and kills thousands annually. Still, when, in June 2009, WHO declared that the spread of H1N1 was a pandemic, some of us were less than alarmed. When virologists were able to rapidly incorporate the novel virus into the annual flu vaccination package, I actively supported vaccination programs. My general view of vaccinations is that they are there more to protect others than to save myself. Whatever side effects the vaccine might incur were part of my commitment to keeping the people around me and in the community healthy. I believed then—as I still do—that many anti-vaxxers are driven by selfishness. As an admittedly privileged, and sometimes smug, white male Canadian, this "me-me-me" business was not something to which I aspired. Still, I had my moments of cynicism. Even as WHO explained pandemics to the general public, and rolled out their six-phase pandemic preparedness plan, I devised my own version of the phases of what I called pandemic panic. My categories started at a pre-panic Phase One, a kind of who-cares phase, in which infection was transmitted among animals in poor countries, then moved into an early panic phase, characterized by moral outrage, in which European and North American tourists returned home with stories about sick dogs or children in the streets. The highest levels of panic and depression among European and American officials, who usually lead the charge on responding to these emerging diseases, involved sick wealthy white people and corporate financial losses on the stock market.

By August 2010, when the pandemic was declared over, and WHO announced that H1N1 had settled into the "typical seasonal patterns," the new virus had officially killed 18,500 people. The mathematical models used to follow seasonal influenza suggested actual deaths from H1N1 ranged between about 150,000 and 575,000. These numbers are not those we might find with Ebola, but they are not trivial.

On the face of it, the "why" questions of the H1N1 pandemic appeared, at first, to be straightforward. The initial explanation was that the virus originated in commercially raised pigs in Mexico. Later, based on comparisons between the genetic composition of the 2009 virus and other swine influenza viruses circulating in the region, researchers discovered that the movement of viruses from Eurasia and the United States into Mexico closely followed the direction of the global trade of live swine.

Pigs were raised in large-scale operations in Mexico for the North American market for reasons similar to those used for large-scale poultry production. Mexico (and many other countries) offered somewhat relaxed labor laws and low-cost workers. An infected farm worker, receiving low wages and with no paid sick leave or health insurance, would have every incentive not to stay home even when seriously ill. Under these inequitable social and labor conditions, the virus would have been transmitted back into the pigs in a positive feedback cycle. An obvious measure to prevent the spread of the virus would have been for all countries importing pigs from these farms to require them to have paid sick leave and health insurance for their workers, and to meet minimum standards of hygiene.

In one sense, every death is expected and tragic. Optimism about technological breakthroughs notwithstanding, death is, and will remain, the usual course of events. The world could not unfold as it should if we did not all, sooner or later, die. This eventuality is what gives us despair even as it opens the possibility of a better future for our grandchildren. Those of us of colonial descent, however, have come to expect that our deaths will occur later in life, by heart attacks, cancer, suicide, or car accidents, not viruses spread by chickens, pigs, or bats, for goodness' sake.

There may be a measure of poetic justice in the idea that a global pandemic could come from chickens or bats. However, what we should be wary of is the pontifications and strutting of half-cocked experts and autocrats, like those who hijacked the global narrative after 9/11. We who are citizens of industrial societies have been the ones who called for a chicken in every pot, using every pork-barrel political maneuver at our disposal. In our conquest of hunger and pursuit of obesity, we were the ones who pushed for economies of scale in agriculture, which not only brought down the consumer price of good protein but also created vast brewing vats for a multitude of infectious microbes. Chickens and pigs, the two fastest-growing livestock populations in the history of the world, have been on the front line of our battles against nature and hunger—and our fanatical obsession with personal health, low taxes, and the economics of me-ism.

This, then, is the revenge of the pawns; the foods we eat are the frontline messengers from a beleaguered biosphere to cocky human beings. So far, at least with regard to avian and swine influenzas, the general response has been to centralize control and depopulate the barns, which is akin to shooting the messengers. If we want to stay around in a convivial world, our species shall have to do better; we shall have to come to some accommodation with the kings and queens and bishops hiding in the back row. This is not a battle to win. The best we can hope for is an uneasy, mutually respectful, mutually wary conversation.

(3)

ZOONOSES AND DISEASES PEOPLE GET FROM OTHER ANIMALS

COVID-19, SARS, and influenza belong to a very large extended family of diseases that are shared among other animals and people. They are called zoonotic diseases, or "zoonoses" (pronounced zoó-uh-no-sees) for short.

Most of the infectious diseases people have ever cared about have come to us from other animals. These include many of the big killers, such as measles and smallpox. A few, such as tuberculosis and its mycobacterial relatives, may have started in the environment and infected humans, after which we infected our animals, and then they transmitted the infections back to us. Some infections made the jump to people, redecorated our microbiome, and decided they liked our lifestyles. They liked what we ate, where we hiked, how we loved to crowd together, how we enjoyed inserting juicy appendages into each other. Other diseases have visited us from their animal hosts fairly regularly, but really, when push comes to shove, they like their old animal homes best. The disease-causing agents that live in non-human animals and are transferred to people, but never actually feel sufficiently at home to be passed on from one person to another, are technically the only true zoonoses. Rabies is the example most people will recognize, or, for the older farmers among us, brucellosis, which also responds to the name undulant fever and was once called Malta fever by British troops.

The microbes that scare us, like SARS-COV, SARS-COV-2, and Ebola virus, are the ones that make an unexpected leap out of a dark corner of the global basement. They scare us in part because we are unsure if they are serial killers, or naughty children, or ghosts in whom we don't believe. We refer to the illnesses these agents cause as emerging infectious diseases (EIDS), as if they are somehow coming out of something, a womb, perhaps, or a deep sleep. The agents that cause these diseases are the spies from the animal world sent to check out human beings as possible new hosts.

One of the many reasons EIDS have caused so much anxiety in public health workers is that they are surrounded with so much uncertainty. Unlike our pre-scientific ancestors, we have come to expect evidence-based certainty. Now, we are unsure whether these agents will become permanent residents in human bodies and populations or whether they are just tourists, infections visiting us on a holiday. All of us would prefer, of course, that after a few weeks of pillaging, they return home, like the Vikings, or drunken college students. That possibility requires that there are some populations of animals around for them to go back to. I have friends who cannot go back to Yugoslavia; Yugoslavia does not exist. And there are infectious agents that might not be able to go back to the animal homes from whence they came, because we are driving those natural hosts to extinction.

After waves of West Nile virus disease, Lyme disease, and food-borne infections with *Salmonella* and *E. coli*, many citizens of industrialized countries have come to accept the idea that infectious agents from animals can make the jump to people. What people don't like are surprises. In 2009, partly in response to the emergence of bird flu, the United States Agency for International

Development (USAID) launched a multi-million-dollar Emerging Pandemic Threats program, which was designed to predict, prevent, and respond to these threats. After all this time, shouldn't we be able to predict where the next diseases will emerge, and manage them when they do? Don't the CEOs of some of the major corporate foundations talk about eradicating diseases?

In late 2019, just before COVID-19 appeared on the horizon, not having seen any major pandemic threats, budget-minded politicians canceled the program. That's the problem with public health programs: if they are successful, not much happens.

Zoonoses have been around for millions of years and will be for millennia to come, our species should live so long. Microbes of all sorts, all those microscopic "animalcules" that the Dutch scientist Anton van Leeuwenhoek first saw through his microscope in the late 1600s, are all around us, in our mouths, on our genitals, on our hands. And certainly all over those other animals with which we share the planet. Your dog could be carrying them, or your cat, or the mosquitoes zinging around your ear as you read this book on the veranda in the evening; they could be in your salad or in those cute raccoons waddling across the backyard. The diseases caused by these microbes have names that are both familiar and odd, and include the plague, West Nile virus infection, Lyme disease, sleeping sickness, salmonellosis, poker players' pneumonia, and tuberculosis. These diseases can—and do—strike many people every day. They show up as worms in the eye and cysts in the belly, chronic heart disease, and debilitating infections of the brain.

Usually zoonoses drive through our neighborhoods and check us out without our even knowing. We then, as some researchers have said, "suffer" from subclinical disease. Not all infections (when the bug gets into you and multiplies) result in clinical disease. Most often they cause a passing, "flu-like" illness. But sometimes they do kill.

The more perceptive among my readers will have noticed a little trick I have been playing in the last few paragraphs. I have been slipping back and forth between diseases and infections, and between animals and people. When we speak of pandemics of diseases people get from animals, we need to be more careful. It is possible to have a pandemic infection in animals or people without a lot of disease. Infection does not equal disease. Infection in animals does not equal disease in people. Thus, we would prefer that if there is a pandemic, it be a pandemic of infection in animals (which avian influenza already is), or at least a pandemic disease in birds (which it already is), but not a pandemic disease in people.

Some writers differentiate these by using the terms "epizootic" and "panzootic" to refer to epidemics and pandemics in animals other than people. This differentiation can be useful. For instance, once a virus has adapted to humans and travels easily between us, it's important to have a language that helps us develop vaccines and appropriate response methods. Outside of the medical front lines, however, where most of us live, this differentiation tends to obscure the fact that people are animals—exceptional animals, to be sure, but animals nonetheless—and that the sharing of microbes among species is normal.

While acknowledging the usefulness of these categories, especially for frontline medical management, I prefer to look at the details of the diseases we share with other animals, or our shared environment, whether or not they spread around the world as pandemics or ride along with mosquitoes and ticks. In the long run, our understanding of the complex relationships among microbial populations, non-human animals, and the habitats we occupy will provide the details that help us prevent or manage the next emergence of disease.

No realistic understanding of zoonoses—and, I would argue, of any human diseases—can be achieved without the contributions of natural scientists (ecologists, entomologists, and zoologists, for instance), health scientists (veterinarians and physicians), historians, anthropologists, social scientists, artists, and just plain folks. As I have said before, and will repeat as often as necessary: we are each other's peripheral vision.

The bad news is that many of the threats of disease we face today are connected to each other—and to many of the things we have come to cherish in our modern lifestyles. The way we deal with one disease, or even one so-called problem, can have far-flung consequences for what happens to other diseases and other problems.

The good news equally is that many of the threats of disease we face today are connected to each other—and to many of the things we have come to cherish in our modern lifestyles. The way that one disease or problem is dealt with can have far-flung consequences for what happens to other diseases and other problems. As Douglas Adams realized when he wrote *The Hitchhiker's Guide to the Galaxy*, the answer to the question of "life, the universe and everything else" is relatively simple and hard to get our heads—and our lives—around. It may not be "42," as the computer in the novel suggests, but it is not necessarily billions of dollars of better biomedical technology and vaccines, as others propose. Some of our other options will become apparent in the tales in this book.

WHO defines zoonoses as those diseases caused by agents that are naturally transmitted between other vertebrates and humans. All zoonoses have what are called natural reservoirs. Reservoirs for zoonoses are the natural homes of the viruses, bacteria, and parasites—those animals or combination of animals and ecosystems in which the infectious agents usually live and multiply and on which they depend for survival. If the reservoir is a single animal, it's called a reservoir host (which should mean, if life were fair, that we would call the microbe a guest, which we don't;

we denigrate it by calling it a parasite or pathogen). Microbes usually don't kill their reservoir hosts, or, if they do, they take a while to do so, or they find ways to get themselves transmitted to another animal (through biting or contaminating the environment, for instance) before the first animal dies.

Most often, humans are "accidental" hosts of the agent (as if in other types of diseases we are deliberately targeted). If we are "accidental hosts," the microbe doesn't need us; we just happen to provide a handy bed-and-breakfast while it is looking for someone else. Sometimes we are a dead-end host, a kind of Hotel California from which the agent can check out but never leave (alive, that is).

Having said that, there is a wide variety of ways to define and classify zoonoses. Many classifications are useful, but none should be confused with the reality of the complex world in which we live.

Laboratory scientists create systems and subsystems, like genetics, differentiated according to bacterium, virus, and parasite species. These classifications are most useful for developing tests, vaccines, and drugs.

Veterinarians and zoologists prefer to classify zoonoses according to the animals that are their natural homes, such as chickens, pigs, dogs, rodents, bats, civets, pangolins, or nonhuman primates. This classification helps us to think carefully about the species with which we interact, how we interact with them, and how we might manage those interactions.

Using this classification, we can begin to understand how people provide opportunities for zoonoses by creating habitats for some types of animals (raccoons, rats, and coyotes, for instance, in urban environments, or white-tailed deer in abandoned North American farmland) or by invading the habitats of other animals, such as bats, wild primates, or migrating birds.

Some of our interactions with animals, and transmission routes of zoonoses, are difficult to classify. The interactions with monkeys in Asia, for instance, can be hard to put into one box. When I was in Nepal, working on a parasite that transmits from dogs to people, I sometimes visited the local tourist sites. About four hundred free-ranging rhesus macaques live at the Swayambhunath temple in the Kathmandu Valley of Nepal, interacting with residents and visitors by sharing water, food, and aggressive play. Scientists have only begun to investigate the disease risks associated with such temples, common throughout South and Southeast Asia. Similar problems arise when eco-tourists transmit diseases to gorillas in Rwanda, or vice versa.

In North America, people play with their pet dogs and cats or take them into hospitals in animal-visitation programs, raising similar questions. The stakes are raised if we engage in that most intimate contact of all: eating another species. One theory of the emergence of HIV/AIDS is that people killed and ate non-human primates, from which they picked up the virus. Why people would dine thus—being marginalized, poor, and driven into the forest margins—is part of the larger story, which I hope becomes apparent as you read this book.

Acknowledging that all classifications may be useful but do not represent the whole messy truth, I still think we can learn a great deal by exploring these diseases from different angles. We may build on the animal-based classification of zoonoses by invoking a modified version of ecology, focusing on the cycles of infection that maintain the disease agents in nature and the ways they are communicated between animals, and from other animals to people. This classification system, advocated by veterinarian and zoologist Calvin Schwabe, is the one which shaped my own initial understanding of zoonoses. Schwabe proposed categories he called direct, cyclo-, meta-, and saprozoonoses.

Direct zoonoses, which include many food-borne agents, like *Salmonella* and *E. coli* O157:H7, may be perpetuated in nature by a single vertebrate species, such as birds or cattle; their natural cycles require no invertebrates, and they are transmitted to people directly, through bites or food. This book will not dwell on food-borne zoonoses; those diseases are more thoroughly explored in my book *Food, Sex and Salmonella*.

Cyclozoonoses have maintenance cycles that require more than one vertebrate species but no invertebrates. *Echinococcus multilocularis*, a parasite found in dogs and other canids, is an example.

Metazoonoses require both vertebrates and invertebrates, such as ticks or mosquitoes, to complete their life cycle. The plague, which needs fleas, and American sleeping sickness, which requires triatomine bugs (also called kissing bugs), are metazoonotic diseases.

Finally, Schwabe created a separate category for those zoonoses that didn't fit easily into the other boxes. Saprozoonoses depend on inanimate reservoirs or development sites, such as soil, water, or plants, as well as vertebrate hosts. In this, he was building on the work of the Soviet zoologist Evgeny Pavlovsky (1884–1965), who, in his landmark work *Natural Nidality of Transmissible Diseases*, considered pathogens from an ecological perspective, with each having its own ecological niche.

Pavlovsky was studying plague in Mongolian marmots, but, closer to (my) home, we can look at some of the parasitic diseases we get from our pets, like toxoplasmosis and toxocariasis. These require an external environment to complete their life cycles; their eggs do not become infectious for another animal until they have "ripened" in the environment for days to weeks.

Other saprozoonoses, which aren't zoonoses by WHO's definition, grow in the environment, from which they make people and other animals sick. In 2001, for instance, dogs, cats, harbor seals, porpoises, ferrets, llamas, and people in and around a particular area of Vancouver Island all suffered serious lung and neurological infections from an environmental fungus called *Cryptococcus gattii*. The strain of the fungus that caused the outbreak usually lives in the tropics and subtropics. Some people suggested climate change might have contributed to its spread.

Cholera was once thought of strictly as a disease shared among people. Now it appears that *Vibrio cholerae*, the bacteria that causes the disease, may have a natural home in copepods, which are microscopic aquatic animals. Given the right water conditions—climate-warmed oceans, runoff sewage nutrients—the bacteria start multiplying. So this may be a zoonosis in the broad sense of the word.

Schwabe's classification by animal reservoirs and natural cycles, which has served scientific researchers and public health advocates well for more than a century, formed the basis for the first edition of this book. I am not sure, however, that it offers the best way to think about pandemics and zoonoses in 2020, in the context of globalization, climate change, ecological collapse, species extinctions, human overpopulation, huge economic and political disparities, and unintended consequences of well-meaning interventions. What we want—what we need—is a way to have our thinking, research, and decision-making encompass all of those issues at once.

In writing this second edition of the book, I have found that the old categories kept falling apart. Everything was connected to everything else. Understanding SARS-COV and SARS-COV-2 requires us to think in terms of landscapes and habitats, how we feed ourselves, and the shifting national boundaries that have characterized the last few millennia of human history. In the chapters that follow on specific diseases, I encourage readers to explore the details of human-animal interactions from different angles including (but not only) natural history and biology, social relationships, economics, and ethics.

(4)

PLAGUES, RATS, AND THE PERFECTION OF FLEAS

According to the late singer and songwriter Leonard Cohen, "everybody knows" that a plague is coming. If they don't know, then Laurie Garrett can quickly set them straight; the journalist penned *The Coming Plague*, the remarkable tome about the adventurer-scientists who have investigated infectious diseases over the past century. If there is not a plague on all our houses, there is at least a plague in every neighborhood. If you google "plague," or even restrict yourself to a more scholarly search of references to "plague," you will turn up not just the Black Death (which was probably caused by the bacterium *Yersinia pestis*) and the white plague, but also the plague monkeys, the forgotten plague, the slow plague, the Bombay plague, the ocean plague, duck plague, fowl plague, swine plague, mosquito plague, a plague of sheep, of cattle, of insects, of censorship, the plague year, plague days, plague dogs, and the Nevada mouse plague. Tracing its origins back to a Greek word for misfortune or blow, the word "plague" has been attached to just about anything that people fear or don't like.

This widespread usage of the word "plague" in our society goes deep to a wound in our cultural psyche. None of these plagues, however, are *the* plague that first visited us before all these more modern plagues. The "original" plague is associated with the bacterium *Yersinia pestis*. This is the same plague that in 1994 so alarmed the world that between 200,000 and 300,000 people fled Surat, India, and startled North American television viewers were faced with the sight of white-masked officials inspecting passengers arriving from India at our airports.

People's fears of the plague are not without foundation. Beginning with the Justinian plague (AD 541–544) in Egypt, the plague crested around the world in at least three major pandemics and is at least partly responsible for pushing European history in new, unexpected directions. For instance, the second pandemic, the Black Death, killed off at least a third of medieval Europe—tens of millions of people—in just a few years in the mid-fourteenth century, even as it killed tens of millions of people in Asia and the Middle East. Economic and cultural changes, including peasants' revolts, the rise of capitalism, and a loss of power by the church, all happened in its wake.

That story, and the horrors and wonders associated with it, has been recounted many times and should be essential reading for any thoughtful person in the twenty-first century. All the lessons about ecology, public health, personal hygiene, climate change, compassion, fate, and the dangers of globalization that anyone would wish to learn are there, plain for those who wish to see. This statement is also true for the story of the white plague, which is actually tuberculosis, another disease discussed more fully elsewhere in this volume (and not to be confused with another white plague, which bleaches coral reefs).

The third plague pandemic started in China in 1855 and from there was disseminated around the world. Between 1898 and 1918, more than 12 million people in India died of that plague. When the pandemic subsided, like a great flood it left small, stable pools of infection cycling among rodents, fleas, and the occasional unwary person in uneasy neighborliness on every continent except Australia. In the 1990s, about a dozen countries reported outbreaks of the plague. Most outbreaks occurred in African countries such as Madagascar, Mozambique, and Tanzania, but cases were also reported from Brazil, Ecuador, Peru, Mongolia, Vietnam, China, and the United States. In 1997, an outbreak of pharyngeal (infection at the back of the throat) plague, associated with eating raw camel meat, was reported from Jordan.

The plague is a disease of increasing concern because of the growth of urban slums globally and the unstable weather conditions associated with global warming and El Niño; these lead to changes in habitat for rodents in the countryside and explosions in rodent populations in urban centers. We tend to view this disease as one that affects people, which, of course, it does. But we may learn something by taking a broader view. The plague is a disease of people, of fleas, of rodents, and, finally, of whole social-ecological systems.

In people, it comes in three forms: bubonic, pneumonic, and septicemic. To cause disease, the causative organism, *Yersinia pestis,* needs to get into the bloodstream, through an open wound, through transmission from one's afflicted cat (say, by kissing the cat and inhaling the bug), or through a bite from an infected flea that then vomits the bacterium into the bloodstream. Much like the first European invaders feeding off the bounty of the American continents, once in the bloodstream the bacterium travels down the lymph rivers and bloodways, looking for appropriate places to settle and multiply. Peripheral lymph nodes—in the armpits or crotch—seem to be as good a place as any. These nodes swell painfully, and the infected person begins to feel some combination of fever, chills, headache, diarrhea, and constipation, moving on to generalized pain, rapid heartbeat, anxiety (understandably), staggering gait, slurred speech, and prostration. If untreated, 25 to 60 percent of people die. This bubonic form of plague cannot be transmitted from one person to another; an epidemic of bubonic plague depends on the continued presence of rodents and fleas.

In some people the infection moves to the lungs, where it causes pneumonia, which is a kind of adaptation of the organism to modern human society. Once in the lungs, it can be coughed out into the air and transmitted from person to person. At this point, fleas and rats are no longer necessary. This form of the disease is called pneumonic plague. Now, or so conventional wisdom (recently called into question) has it, the bug can cause a serious epidemic or pandemic.

Although both bubonic and pneumonic plague can kill, they don't kill everyone. At least some bodies are able to fight the disease off, and people recover. In some cases, both these forms of the disease lead to a full-blown septicemic (that is, blood-borne) form, in which people develop nervous and cerebral symptoms very quickly. Pretty well all of these people die.

But the plague is not just a disease of people. It is also a disease of fleas, which are themselves something of a wonder and worthy of a small aside here. In the history and mythology of the Judeo-Christian tradition, three and seven are two numbers used to express perfection. The third perfect number in Judeo-Christian stories is forty, which is the number of days Jesus spent in the desert, and which is an excellent age to have been, but which is not relevant to the life of fleas. The flea, aside from never experiencing the joy of reaching that pinnacle of the roller-coaster of life, the fortieth birthday, is a perfect parasite.

Besides being perfect, which in and of itself is annoying, fleas are exceedingly small. They have caused me to quarrel with my wife, whom I accused, no doubt falsely, of hallucinating tiny jumping, biting specks. They are brown, wingless animals with Twiggy-like bodies that can make them seem to disappear when they face you head-on, or more likely, rear-on, and that enable them to slip unpunished into cracks and cleavages. Suctorial mouthparts, eyes, antennae, and combs bedeck the chitinous head, giving them, under the microscope, the warrior-like appearance of a military beast from a planet in the Star Wars galaxy. The thorax is divided into three, which is another sign that the (alleged) Creator, as several biologists have (allegedly) suggested, had an inordinate love of insects. From each segment of the thorax extends a pair of powerful, claw-tipped limbs. If fleas, says a scribble in the margins of my parasitology notes from my student days, were the size of people, they could jump over tall buildings. More relevant to me is the fact that fleas the size of fleas can jump from the pillow on the couch to the base of my ear.

If one were to go to the flea for moral lessons, one could learn that if it drinks enough (blood) and copulates with sufficient frequency, even a small creature can have a significant impact on the world. For those who equate "good" with "natural" and "natural" with all those living things that aren't us (a strange and pernicious equation), this may be a startling insight. Under the right conditions of decadence, a female flea, which lays from three eggs to some much larger number (probably a multiple of three) at a time, can put out a hundred of the glistening-white little things in a one-year lifetime. Fleas will lay more eggs when it is warm and the humidity is 70 percent or more. The adults live on warm-blooded animals and eggs are laid on the animals, but many of them fall off into rugs and carpets, where, like jelly beans and chocolate eggs left unfound under the sofa cushion on Easter morning, they will hatch in two to twelve days (the median of which is seven).

The yellowish-white, bristling worms—teenagers of the insect world—feed on adult flea feces (full of dried blood) hiding out in the carpets and pet-resting places where they were dropped off.

They are graced with anal struts, the flea version of tight jeans. Whereas people use their struts to unhinge each other's psyches, the flea larval struts are hooks used to propel them in search of the ultimate hamburger (fecal blood casts from the adults) to stuff into their chewing mouthparts. They feed, grow, and molt three times, a process that can take anywhere from nine (three times three, under perfect conditions) to two hundred (or maybe two hundred and forty-three, which is a multiple of three) days. After the third molt, the larva spins a small cocoon, within which, like a most excellent pupil of the natural order, it pupates from seven days (under perfect conditions) to one year (usually two to four weeks, the median of which is three), after which the now-mature adult flea breaks out, seeking a host on which to feed.

In flea heaven, which has on occasion been situated on our couch, the entire life cycle takes three weeks. But fleas are quite willing to accommodate themselves to our imperfections and will stretch the process out, if they must, to two years.

There are about three thousand species and subspecies of fleas, of which only a few infest us and our pets. *Pulex irritans*, the human flea, moves freely between dogs and people, especially in the southern United States. This flea can transmit murine (mouse-related) typhus, which has less catastrophic effects than epidemic typhus, which is transmitted from rats to people by lice and has caused massive suffering and death worldwide. Epidemic typhus has been the subject of one of the classic biographies of a disease, *Rats, Lice and History*, by Hans Zinsser, which should be in every intelligent person's library.

Ctenocephalides felis (the cat flea) and *Ctenocephalides canis* (the dog flea) infest dogs, cats, and people in a kind of natural free-trade zone, although the tapeworms that they sometimes smuggle across the borders only make a home in dogs. The cat flea is probably the most common flea infesting North Americans and is the most difficult to get rid of. *Echidnophaga gallinacea*, the sticktight flea of poultry, is nature's punishment for wayward dogs, cats, and young children (myself, many years ago, included) who decide to check out the inside of the chicken coop. In cats and dogs, this flea digs into the sparsely haired regions around the eyes, between the toes, and around the ears; in people, it just looks for uncovered, hairless areas.

Until relatively recently, there were few good remedies for getting rid of fleas, which spawned a huge industry in equally (in)effective treatments. Substances that effectively killed adult fleas (apart from the suffocating effects of baths) were toxic not only to fleas but also, because we share the animal kingdom with them, to us and to our pets. A California company proposed sealing off and heat-treating houses to get rid of insect infestations; this kind of drastic solution should be implemented just after first frost, in those parts of the country that get a first frost.

Other remedies for fleas proposed by several authors include: feeding the afflicted animal brewer's yeast (a teaspoon daily for a cat, a tablespoon for a fifty-pound dog), and garlic (one or two cloves per day); regularly bathing the animal with non-toxic shampoos (to suffocate the fleas), followed by a good brushing; and rubbing the dog or cat with eucalyptus oil in water and/or brewer's yeast, especially over the back around the base of the tail and in the neck area. Because treatments of afflicted animals kill only the few fleas unlucky enough to drown or get squished, baths and powders should be applied outside, where the fleas do not have the opportunity to escape into your carpet or couch. The standard veterinary literature reports that no controlled trials have demonstrated that these proposed treatments are effective. Even the herbal-treatment guru Juliette de Bairacli Levy somewhat despairingly admitted that her protective lotion against ticks and lice, containing wormwood, beer, and vodka (a combination that would have a profound effect on many of us) does not affect the spartan flea.

In recent years, various hormonal remedies that interrupt larval development, combined with drugs that kill the adults, seem to be effective and safe. Removing fleas from your pet and discouraging them from reinfesting is only symptomatic treatment, however, and may aggravate the problem. Remember that many of the eggs and larvae are just hanging around the house, waiting for you to sit down. Cat fleas in particular are not fussy eaters, and, unless your whole family is on brewer's yeast, garlic, and eucalyptus baths, sitting on the couch may become a bloodbath. If you are truly serious about fleas, you must remember that though you treat your dogs or cats with baths and powders, though you deck them in eucalyptus collars and feed them garlic, brewer's yeast, or thiamine, yea, even though you spray them with the most powerful toxins, causing them to suffer even to the point of death, and treat not the couch and the rug, what you have done may not bring you to flea-free heaven.

Cat and dog fleas, so I am told, do not live well at altitudes higher than five thousand feet, so if you find that you cannot live with fleas, and you cannot, for reasons of conscience, live without them, you should lift your eyes to the hills. Know also, however, that even if you escape to Nepal or Tibet or Colorado, the tiny rat flea, *Leptopsylla segnis,* survivor in cold, dry climes, may await you there. There are fleas living on every continent, including Antarctica. Someone found a 200-million-year-old fossil flea in Australia.

There is no escape from the perfection of nature. It has predated us and shall outlive us. The religious among us might wonder if the creator invented the plague to make fleas sick and die and if we are the accidental hosts. If so, then getting rid of the plague may only encourage fleas.

Which brings us back to the plague.

Scientists suspect that *Yersinia pestis,* through one genetic mutation, adapted itself to transmission by flea bites thousands of years ago. Its cousin *Yersinia pseudotuberculosis,* which lives in rats and pigs and causes appendicitis-like food-borne illness in people, never made this adaptation. Fleas pick up *Yersinia pestis* from the blood of infected animals; the bacteria then multiply in the flea's stomach until its upper gut (proventriculus) is blocked up. Really hungry and sick to the stomach, the flea heads out to a local blood bar, ready to feed on anything—your leg, a dog's tail, a rat's butt. After taking in a huge gulp of blood, the flea regurgitates some of the bacteria back into the glass (that is, the wound on your leg), then takes a crap into it and heads on home.

A lot of fleas probably feel miserable at getting this food-borne disease (the hamburger disease of the flea world), and some of them die. Even people who love all living things usually don't care much about the personal sufferings of fleas, real or imagined, however, and no therapies have been developed to treat them and make them feel better. Penicillin would work, if one could devise a way to deliver the drug. As a veterinarian, I have had enough trouble trying to inject skinny cats and those small, nervous, snappy, skin-and-bone dogs without trying to imagine how one might inject a flea. Still, the nanotech people, I am sure, could find a way, should we decide to be more generous to other species than our reputation would suggest. One might also try antidepressants, since they have worked in a wide range of other species; it wouldn't cure them, but they would feel better while they died, which is, I suppose, the best most of us can hope for anyway.

If fleas are bearers of bad plague news, then our warm-blooded relatives the rodents are the poor suckers who have fed and housed them all these many years. Fleas had to wait a long time before getting a good meal of rat's blood; rodents seem to have appeared on the planet some tens of millions of years ago, probably somewhere in Asia. Rats are even more recent. They appeared a few million years ago. *Yersinia pestis* causes in small rodents diseases similar to those in people, dogs, and cats.

Wild rodents seem to be more resistant to the plague than domestic rodents, although plague has been known to cause major epidemics and die-offs in prairie dogs and other rural rodent cousins. Rats can carry a wide variety of diseases that can be transmitted to people, including plague, leptospirosis, typhus, spotted fever, tularemia, salmonellosis, and hantaviruses. I'll come back to some of these a bit later.

According to Hans Zinsser in *Rats, Lice and History,* black rats (*Rattus rattus*) arrived in Europe from the Far East sometime between AD 400 and AD 100. They may have been fleeing the brown rats (*Rattus norvegicus*) that ruled Asia. They may have meant to warn the Europeans about their brown relatives. The Europeans didn't listen. The so-called Norwegian rats (*Rattus norvegicus,* carried to Europe on Norwegian boats) followed from northern or northeastern China about a millennium after the black rats and pretty well wiped out their cousins.

Norway rats are Darwinian wonders, following and adapting to human habitats and adopting human behaviors with a vengeance. They reproduce with biblical perfection—about seven litters a year with seven pups per litter (well, with variance around those numbers)—and the females can breed eighteen hours after giving birth. This fecundity enables them to bounce back quickly from disease, pestilence, and human attacks. They are omnivorous and have been known to eat a third of their weight (another good biblical number) daily, voraciously consuming garbage, soap, candy, fruits, grains, seeds, and other animals, including each other. They live to be about three years old, so they live fast. They kill anything that gets in their way. They have studied us well, and bettered us.

Having been born in the Chinese Year of the Rat, I confess to having some empathy for the creatures. A group of rats is called, among many other, less flattering, names, a mischief. Researchers have determined that rats are not only smart but loyal, rescuing mischief mates who are in trouble. They are also, apparently, ticklish. Still, I am troubled by a case described to me by a colleague who, having just returned from Tanzania, was called upon to perform some eighty dollars' worth of medical and surgical procedures on an urban pet rat. That's one of many reasons I am happy not to be in private veterinary practice. Rats, in my humble opinion, are intended to have honest, hard-working lives in neighborhoods other than my own.

No one knows how many rats there are in the world, although it would not be wildly outrageous to say that there are as many of them as there are of us. Although we are never, it seems, far from rodents, most residents of North America and Europe have had few encounters with them. Growing up in a clean prairie Canadian city, I never had much to do with rodents. Reports in 1998 of rats rampaging through Chinese cities, unhindered by feral cats and dogs, chewing through garments of hotel guests and damaging power lines, do not ring any bells from my own childhood. I imagine that if such events had happened, I would have suppressed the memory so deeply that only radical psychotherapy could retrieve it.

I once caught a mouse as it shambled along the street, close to the curb. It was Sunday morning, after church, and the sun was hot: a usual Winnipeg summer Sunday. The mouse, resisting my evangelical attentions, bit my finger when I grabbed its tail. My parents, if I remember rightly, took me to a doctor. There was some concern about tetanus. Nobody mentioned diseases such as plague or hantavirus pulmonary syndrome, although we now know the agents that cause them are present in western Canadian rodents. And once, when we came home from Sunday evening church services, I remember a lot of squealing and excitement in our family; there was a mouse in the mousetrap in the front hall. The concern in this case, I think, was about protecting our food supplies.

In the southwestern United States, ground squirrels can carry the plague, which has made for at least one exciting, if maudlin, made-for-television movie. In 2006, a woman in Los Angeles picked up bubonic plague from an unknown source, probably a flea that jumped from a rodent to a cat to a person. The occasional cat has passed the disease directly to an overly affectionate owner by rubbing noses.

I suppose if I had grown up on a prairie farm, I might have learned about plagues of gophers. In this case, the gophers themselves were the plague, wreaking their damage on fields. Once, in high school, I went out to Eddy Wall's farm, and we walked around with twenty-two-caliber rifles and looked for gophers to shoot. I don't remember seeing any, though I shot a few tin cans off a fence post. At that time, no one knew that small prairie rodents around the world carry dozens of different kinds of bacteria, some of which (*Bartonella* species) have, on rare occasions, infected people and caused fevers or heart problems.

My first serious encounter with rodents was in Bihar, India. At nineteen years of age, I had hitchhiked overland from Europe to Nepal and India, where I worked as a young, idealistic volunteer. Actually, that is a romantic oversimplification; running out of money in Calcutta, stunned by the poverty there and my lack of money, I had talked my way into a temporary job with the Mennonite Central Committee, helping with grain storage and distribution in a grain-for-work program to build earthen dams and improve agricultural production.

My first day in Bihar, I slept on the floor of a bookshop, and some rats brushed against my cheek in the night. Several months later, I was part of a team effort to round up rats from a grain storage building; somebody made a lot of noise behind the stacks of grain sacks, and the rest of us waited for the rats to come out. We were supposed to kill them. One rat (the only one I remember, though I know there were a lot of them) came running out and skittered up the wide pant leg of one of the Indian workers, over the crotch, and down the other leg. At least, that was the story. Lots of excitement about possible theft of the family jewels, but no talk of plague.

I don't remember our killing any rats at all; I wonder in retrospect whether our Indian co-workers, coming from a culture in which some of their compatriots feed and worship rats as companions of the god Ganesh, were just putting on a charade for the benefit of squeamish (and wealthy) westerners. They could have tried to explain things to us, but we, whose competitive and jealous God has no animal companions, would not have understood. Better to let us play our little games. Before I left Bihar, I acquired a pet chipmunk (another potential plague carrier), which I hauled around in my pocket and which ran up and down my arm to get snacks out of the palm of my hand. Children on the crowded train from Ranchi (Bihar) to Calcutta (now Kolkata) loved it. The chipmunk later died after being accidentally locked in a refrigerator.

I tell you these stories not because I expect you to be interested in my experiences with rodents but because they indicate how far removed many of us in the industrialized world have become from the terrors of the Black Death. Yet the world would be a very different place if the plague had not turned Europe into a post-nuclear-like disaster area in the mid-fourteenth century. We are children of the plague not just in our language but also in many of our public institutions and political ideas: the collapse of feudalism, so I understand from historians, owed more to the deaths of farm laborers from the plague than it did to our enlightened moral progress as a species. And this collapse of feudalism in western Europe led to the rise of the middle class, capitalism, shopping malls, and evangelical consumer cults.

Although most people think of the plague as a disease of people, and some of us (veterinarians) consider its effects on other animals, few of us think about the fact that the plague is also, perhaps most accurately, a disease of socio-ecological systems.

The Black Death occurred because the Mongol Empire had created a network for transport of goods, services, military technology, rats, fleas, and bacteria from central Asia to the gates of Europe. The Black Death occurred because Europeans, responding to longer summers and shorter winters between 700 and 1250, tripled their population, exhausted their agricultural lands, and were unprepared to adapt to the mini–ice age that followed. The Black Death occurred because prevailing winds changed, making Europe wetter—and central Asia drier—than it had been for a few hundred years. Mongol and Turk nomads, herding their flocks and joined by marmots, susliks, and other types of wild rodents, headed east and west in search of greener pastures. Europeans starved to death by the millions before the plague reached them, and those who survived the famines were weakened when the plague arrived. In this version of the narrative, *Yersinia pestis* may not even be important.

One author has suggested that the clinical signs and recurrent epidemics of the Black Death in Europe were more likely anthrax—another fatal disease people share with animals—than plague. Between outbreaks, anthrax can survive in spore form for decades and perhaps centuries in the soil (or in laboratories, but that's a different story). What is important in this narrative is that the bacteria that killed individuals were opportunists taking advantage of climatic, environmental, socio-political, and economic changes. If the plague bacteria hadn't done the job of killing, some other organisms would have; there are always plenty of candidates. And this is true for many human diseases: the specific agents (bacteria, viruses, prions) are less important than the hospitable contexts people create for them.

In 1993, a severe earthquake rattled parts of central India, damaging rodent habitats and resulting in an explosive emigration of rats from their own neighborhoods into nearby villages, where damaged grain storage facilities provided easy food and shelter. Wild rodents, under normal conditions of easy country life, may carry *Yersinia pestis* without getting sick. This phenomenon is not usually true of city rats or of rats living under crowded slum conditions. One could make a political lesson from this, but it seems too obvious.

About a year later, in August 1994, some people noticed a die-off in the rat population; after thirty years of no reported cases in India, the plague had arrived. Within weeks, some eighty cases of bubonic plague in people were reported. Since bubonic plague isn't transferred from person to person, health authorities treated the sick people and sprayed infested houses. Nobody died.

In late September, cases of pneumonic plague, which can be transmitted from person to person, showed up in Surat, a city of 1.4 million to the northwest of the original outbreak. When these cases were reported, more than 200,000 people headed out of town, and the plague became world news, with images of airport officials in Europe and North America wearing face masks and carefully screening airplanes from India. This was pandemic fear before SARS, COVID-19, and avian influenza. But the pandemic itself never materialized. Some public health officials now wonder if the disease was plague at all.

Outbreaks of pneumonic plague among miners in the Democratic Republic of the Congo in 2005 and again in 2006 sickened thousands and killed hundreds, but they did not presage a pandemic explosion, as some had feared they would. Some scientists began to have second thoughts. Had *Yersinia pestis* changed its genetic colors? Were nutritional status and living conditions in the twentieth century, even at their worst, far better than living conditions a hundred years ago?

In Uganda, in 2004, a policeman from the West Nile region (ah, that cradle of Caucasians' darkest fears) called a physician, who called staff from the CDC who happened to be in the area for a field trial on treating the plague. They found that two people had come down with pneumonic plague, and despite many human contacts, only two others—their closest caregivers—picked it up from them. You had to be close enough to get sprayed by the coughing. So, people with pneumonic plague should stay off the stage at Stratford (a prime center of spraying others with well-articulated Shakespeare). But maybe our fears are larger than the threat. Fear of our own stories, our own perceived history, casts a long shadow across our cultures.

Several years ago, one of my colleagues, an anthropologist involved in international research, came to me with a problem. "Listen," she said, "I'm working with some very dedicated scientists in a part of the world where plague has become endemic. Take a look at these reports, which concern one particular village. You're the ecosystem health expert. What should these people do?"

From the unofficial research project reports she handed to me, and a review of two decades' worth of peer-reviewed literature, I pieced together the following story:

About a century ago, the plague arrived in an area of what is now Tanzania along the trade routes and became established in several parts of the country, but not in one particular village and its environs. Then, several decades ago, the inhabitants of this village were plagued with an explosion of the rodent population, but without the occurrence of the plague. Nevertheless, the damage from the rats was sufficient to require the intervention of a special healer, who, in the manner of any good epidemiologist, intervened when the epidemic was at its peak and thus would decrease no matter what he did.

A couple of decades later, the rat plague returned, this time carrying human disease with it. Local people consulted the healer, who, it turns out, had not been paid the last time; after receiving appropriate apologies and offerings, he again suggested some cures. This time, however, perhaps because the people didn't believe hard enough or do exactly the right things, the plague persisted. Over the decade of the 1980s, thousands of people became sick, and hundreds died.

In the meantime, veterinary scientists were trapping rats and taking blood and fleas from a random sample of people and dogs, looking for *Yersinia pestis* or antibodies to it. Every possible preventive action seemed to have been tried, but nothing worked, at least not very well. Some research suggested that dogs were carrying the disease; maybe if they got rid of the dogs? An anthropologist was sent in to talk to the people to see if he could determine why nothing seemed to work.

A complicated picture emerged from the anthropologist's report, including what had been tried and why it hadn't worked.

Doctors tried to quarantine sick people, but the patients resented this intrusion into their daily lives. Under quarantine, they were prevented from leaving their homes or villages to take part in agricultural activities, gather food, care for sick relatives, or attend special occasions; some people could find loopholes, however, especially through bribery. In any case, if rat fleas were carrying the disease, it was not clear that quarantine would have any effect besides making it look as if something were being done.

One could fall back on simply diagnosing and treating cases as they came up. Since the plague is generally treatable with inexpensive antibiotics such as tetracycline or streptomycin, this approach is often the most cost-effective for sporadic (non-epidemic) cases of the plague. Treating people quickly is dependent on local treatment centers that are accessible, well staffed, and stocked with antibiotics, however. In Tanzania, the local treatment centers were accessible but not always stocked with drugs; patient records were sometimes mixed up, lost, or leaked beyond the center, raising questions of confidentiality. Patients were to bring their own bedsheets and the like; some could not do so and were ashamed of their poverty.

In many parts of the world, diagnoses are made by playing the odds. “If you hear hoofbeats,” our professors at veterinary school in Saskatchewan used to say, “think horses, not zebras.” But that advice needs to be taken in context. In Kenya, you might well think “zebra” at the sound of hoofbeats. In North America, we often call a fever together with either respiratory or gut problems the flu; but often it’s not influenza (especially the gut form, since influenza in people is usually a respiratory disease). Sometimes that doesn’t matter; “flu-like” illnesses in North America are often caused by viruses, and doctors usually treat “flu-like” viral diseases similarly (chicken soup and bed rest). In parts of Africa where the plague is known to be endemic, a lot of non-plague diseases are called the plague, including malaria. Unfortunately, malaria is not treatable by antibiotics. If only some forms of the plague, broadly interpreted as any general illness, responded to antibiotic treatment, then what advantage did those modern methods have over traditional healing?

Public health authorities tried pesticide spraying and rat poisoning in houses. People from households that had been sprayed were asked to shut all windows and doors for six hours, as well as not to clean house for three to six days. Those householders did not like to share a dwelling with dead rats and fleas for that amount of time. Moreover, the pesticides had good market value and were either allocated according to favoritism or resold to farmers as treatments for crops and foodstuffs. Some people complained of becoming ill after spraying; chickens sometimes died, and cattle became ill.

Authorities proposed that villagers plaster and seal their houses so that the rats couldn’t get in from outside. Most of this work was supposed to done by women and by children under twelve, who were the ones getting sick and who did not have the energy for house renovations. Besides, there was a shortage of water needed for plastering, and men controlled the money that would pay for plastering. It was not done.

Households could move food storage outside so that the rats wouldn't come into the house: traditionally, the maize was stored outside the house. However, a decline in maize production associated with an increase in cash crop production related to an opening of world markets made maize more valuable as a tradable commodity; this development, together with a decrease in social trust (related in part to competitiveness and modernization), increased the need for vigilance. Hence the maize was now stored in the wooden ceiling of the house. Rats followed the maize into the house; this move made the dogs happy to stay around the house as well.

Public health officials recommended that householders clear shrubs and bushes from around the houses and field crops so that rats would have fewer places to hide. But arable land was scarce, and people wanted to maximize land use by planting very close to the house. Borders of fields were planted with trees to prevent soil erosion, with grasses as cattle feed, and with medicinal plants for household use. These practices were encouraged by some government agencies (obviously not those involved in eradicating the plague) to conserve soil and water. Some shrubbery was also maintained as a link to family ancestors, who were believed to live in the shrubs near the house.

Even something as apparently simple as removing food scraps from near dwellings was a problem, since this organic "waste" was thrown into the fields around the house to serve as fertilizer for the crops.

Since women and young girls prepared the food, they were getting the plague. Since children played with dogs that were infected by fleas from rats, they also got the plague.

What should these people do? To keep up agricultural production and prevent soil erosion, farmers should plant bushes and shrubs around their houses and fields. To prevent the plague, they should pull them up. To make more money, they should plant more cash crops. To bring down the price of local food, they should plant more maize. To improve their health, they should co-operate with their neighbors and move their grain into outside storehouses. To become more competitively efficient in the global market, they should compete with their neighbors. How could they decide the right path? Was there a scientific answer?

When I recounted this scenario to Jerry Ravetz, a philosopher of science and a colleague, he suggested (not without some satisfaction, I think, since philosophers, like mathematicians and theologians, seem to thrive on insoluble problems) that there might be no solution to the plague problem in Tanzania. We were up against irreducible complexity.

One might also, on a world scale, say we were up against some major cognitive dissonance. When I became a veterinarian, I was introduced to the idea that rats are intelligent, friendly, social animals that make ideal pets, especially for children. They can do tricks. Hence the eighty-dollar vet bill for the rat I mentioned earlier. Personally, I get a headache trying to reconcile rat worship, rat companionship, and the ravages of the plague. One might suggest that we have need of a kind of eco-narrative therapy for our dysfunctional species, but where would we begin?

Clearly the plague, once the terror of Europe, is no longer the killer that it once was. What did Europeans do correctly that the Tanzanian villagers can learn from?

For one thing, it has become clear that no single perspective (environmental, public health, economic, male, female) or scale (animal, household, region, world) will give us the whole picture or, if there is one, the whole solution. If the plague is seen to be a disease of individuals caused by *Yersinia pestis,* then the treatment is an antibiotic, and the delivery of disease care is where time and effort should be focused; better diagnostic laboratories need to give people greater confidence that they are being treated by the right drug for the right disease. If the plague is viewed as a disease of household dogs, rats, and fleas, then efforts should target controlling dogs, spraying for fleas, killing rats, and making sure that people have better housing. Land use planners and ecologists can monitor ecological change and target interventions when they are likely to have the greatest effects. These measures, too, are important.

The plague is also a disease of socio-ecological systems. The price of maize, the national debt, relations between men and women in the household, the sense of neighborliness between households all need to be considered. Addressing these issues requires not just medical and technical expertise but also a deep cultural engagement and a democratic debate about trade-offs: what kind of a society do we want to live in?

Similarly, outbreaks of the plague in Mexico, Peru, Madagascar, and other countries in the 1990s could be attributed to a mix of bacterial nests left over in various landscapes from the pandemic a hundred years earlier, as well as to a lack of public health infrastructure (reflecting government spending priorities, which often reflect the muddle-headed priorities in the World Bank and the International Monetary Fund) and to cyclic meteorological events (El Niño), which have been intensified by global warming, which is the result of fossil fuel use, which is associated with industrialization in Europe and North America.

Although I am not an alarmist with apocalyptic visions of the Black Death wiping out modern civilization, I do know that the stories that joust for control of national policies today—tales of competitiveness and economic efficiency, economies of scale and unfettered trade—are also tales of sloppy biological thinking. The word "plague," according to my best epidemiological models and prognostications, shall remain an essential ingredient for intelligent discourse for many centuries to come.

The stories we choose, and the responses to disease, are not scientific choices. They are moral animals, with scientific organs and possibly tragic consequences. If we continue to act fast and loose with our technology and drugs, we may be left without a technology to save us and with the words of Albert Camus ringing in our ears:

> None the less, he knew that the tale he had to tell could not be one of final victory. It could only be the record of what had had to be done, and what assuredly would have to be done again in the never ending fight against terror and its relentless onslaughts, despite their personal afflictions, by all who, while unable to be saints but refusing to bow down to pestilences, strive their utmost to be healers.
>
> ALBERT CAMUS, *The Plague*

(5)

LYME, WITH A TWIST

FLEAS ARE not the only arthropods to have teamed up with small rodents with the sole purpose, it seems, of killing or annoying people; nor is the plague the only disease to have taken advantage of this multispecies collaboration to make the leap from other animals to people. Lyme disease has few of the terrifying overtones of the plague, and not a few people have confused the issue by thinking Lyme referred to a fruit and not to a town in Connecticut. But this emerging zoonosis, and the ticks associated with it, have much to teach us.

Many of us who recall pictures in grade school textbooks, or who have gazed dizzily through university microscopes, tend to think of bacteria as tiny spherical or sausage-shaped creatures that multiply into grape-like clusters or strings of wieners. They multiply fast, more orgiastically multiplicative than the most virile polygamous rabbit you can imagine. One of my friends, a food microbiologist, described to me how one clan of bacteria, *Clostridium perfringens,* can double its population every eight to ten minutes, producing over a million offspring in four hours, and, barring food, water, and temperature constraints, could multiply to fill the entire planet in about forty-eight hours.

Many people think of these little beings as hanging out in bad places, such as slums, poor countries, unbrushed mouth cavities, and insufficiently sprayed kitchen counters. Bacteria are in the air, on your curtains, on the chair, on your food, on your dog, in your underwear. My goodness, how do we even stay alive? Not content with creating colonial blobs at the back of the refrigerator, bacteria not only use animals, people, and hamburgers as public transit vehicles, but some have even developed their own wheels, so to speak.

According to the microbial ecologist Lynn Margulis, very early in evolution some techno-wizards of the bacterial world developed whip-like tails that are spun around by little proton motors. The motors are propelled by changes of electrical charge. The general term for these motorized tails is "undulipodia." In bacteria (like *Salmonella*) and sperm cells (which are probably just bacteria adapted to warm, moist, and pleasurable places), these are called flagella. In your nostrils and trachea, the smaller, waving, hair-like things that push mucus and dead bacterial bodies up the tubes and into the great recycling heap of the biological world are called cilia.

These are all quite interesting. More relevant for our story, however, are those in which the motors were internalized, which took bacteria from having motorbikes to being little sports cars. When the whip-like twisting filaments became internalized, these bacteria, now called spirochetes, were transformed into the fastest-moving dudes of the microcosm. These fast movers took off to newer, warmer, greener pastures, leaving the old, immobile stick-in-the-muds to survive or die at the whims of local conditions.

Margulis hypothesized that when spirochetes joined the human Borg colony, they became the cells that make up our brains and nervous systems and the rods and cones of our eyes. This development means that when we look through a microscope at spirochetes, we are nothing more or less than bacteria looking at bacteria, trying to understand ourselves. Reading this statement, I start to feel a little not-myself, dizzy, and a little, if you please, used. But I suppose human beings don't have much of a leg to stand on when they complain to the cosmic court about feeling used.

Although many of these spirochetes are of interest only to those poor folks like myself who are dumbstruck at the wonderfulness of the world, or to dentists with microscopes spelunking in your mouth, a few are downright bothersome. The one that causes syphilis (*Treponema pallidum*), the dark twin of sperm cells, has been a perennial troublemaker. Two other spirochetes—*Borrelia burgdorferi*, which gives us Lyme disease, and the leptospires, which give us one of the most common zoonotic diseases in the world (not surprisingly called leptospirosis)—are of special interest to zoonotic-disease people. I'll deal with *Borrelia* here and come back to the leptospires in another chapter.

In the mid-1970s, a couple of mothers from Old Lyme, Connecticut, began keeping track of what local physicians had diagnosed as juvenile rheumatoid arthritis. It seemed to these mothers strange that this disease, generally considered rare, should be clustered in their village and in the neighboring villages of Lyme and East Haddam. The persistence of these mothers finally paid off when their story was picked up by Allen Steere, a post-doctoral fellow in rheumatology at the Yale University School of Medicine, who had completed training in epidemiology with the CDC in Atlanta. The subsequent story unfolded through a combination of good luck and doggedly good normal science, as researchers wrote down histories of patients, took blood samples, caught ticks from epidemic and non-epidemic areas, and cultured and tested for everything they could think of.

Willy Burgdorfer, an internationally recognized expert on tick-borne diseases, was actually looking for something else among the crushed remains of a tick on his microscope slide. He was looking for small organisms called rickettsiae, which cause Rocky Mountain spotted fever, the same organism that Herald Cox, at the same institution, was looking for when he cultivated the cause of Q fever (see the chapter on poker players' pneumonia). Instead of rickettsiae, Burgdorfer saw spiral-shaped bacteria.

The spirochetes Burgdorfer saw, in the same general class as those that cause leptospirosis and syphilis, came to be called *Borrelia burgdorferi*. Further testing showed that they were, indeed, the immediate cause of Lyme disease. The medical and scientific community patted itself on the back for discovering what is now thought to be the most common arthropod-borne zoonosis in North America. Nobody likes to talk about the fact that it would not have been discovered if it hadn't been for a few very persistent, annoying mothers, and what this means for how science really works. Few talked about the fact that the Europeans—the Swedes, in particular—had known about the target-rash form of the disease since the early part of the twentieth century and had discovered that it responded well to penicillin. Scientists, I suppose, are as self-absorbed and lacking in humility as anyone else.

We can think about this disease in a variety of ways. Most people, like the mothers in Connecticut, think of it as a disease that affects individual people or animals. That it certainly is. Infection with *Borrelia burgdorferi* (now referred to in scientific papers as borreliosis, further airbrushing the moms from Lyme out of the picture) causes problems not just in individual people. It also affects dogs, horses, and several other species of animals. Several days to a month after the initial injection of the organism from an infected tick about the size of a rice grain or a poppy seed, people may develop a target-shaped rash and some flu-like symptoms such as fever and general pain. The disease seems to respond readily to various antibiotics at this stage. If not treated, the spirochetes swim through the blood and get stuck in tiny capillary laneways, where they cause heart problems, arthritis, neurological problems, and musculoskeletal pain.

Studies by Richard Ostfeld and other ecologists demonstrate that the natural life cycle of *Borrelia burgdorferi* involves field mice, deer, and oak trees. Once every three to four years, oak trees put out a very large acorn crop (called masting). These acorns, rich in fats and proteins, attract field mice and deer that feed on them. The mice also love to feed on the pupae of gypsy moths, a natural pre-emptive strike preventing the moths from attacking the oak tree. Adult ticks, usually some variety of *Ixodes,* drop off the deer where they have been feeding and lay some eggs. The eggs hatch and the larvae feed on the mice, from whom they pick up the spirochete *Borrelia burgdorferi.* After a good blood feed, the larval ticks molt into nymphs, which, the following spring, latch on to whatever fast-food outlet happens to wander into their territory: mice, birds, dogs, hikers, and hunters looking for deer. Late in the summer (or, until recent warming trends, the following year, in some colder places like Canada), the nymphs become adults, and another cycle begins.

From a "rare" affliction in the 1970s, Lyme disease has become the most commonly diagnosed tick-borne disease in North America. Americans now see more than twenty thousand cases per year. The disease has also been moving north. In Canada, between 2009 and 2016, the number of reported cases increased from 144 to 992. Why is this disease spreading, and what, if anything, can be done about it?

Based on a general understanding of the ecology of the disease, a natural history of Lyme disease in North America might go something like this: Early settlers arrived in New England, and, disturbing the natural life cycles of *Borrelia burgdorferi,* mice, trees, deer, and so on, acquired a variety of aches, pains, and diseases, which were the normal accoutrements of being human. They cleared the forest habitat to make way for farms, and the deer populations dwindled, as did the disease and human memory of it. In the 1970s, economic policies based on abstractions created in the ethereal minds of a few academic economists and on the greed of corporate speculators combined with the lure and food needs of big cities to encourage fewer, larger farms (the so-called economies of scale). Thousands of farms were sold or abandoned in New England (and across North America) and often transformed into "country properties."

This economic disaster for farm families was a major success from some other environmental and social perspectives. Suburban "country" dwellers, weaned on Bambi and that family of friendly mice the Littles, decided that nature, especially nature full of non-threatening, picturesque species and devoid of serious predators such as wolf packs and hunters in bright orange vests, was a good thing. As a result, forests made a comeback in New England, along with deer and field mice and a lot of people going for walks in the woods.

All this social and ecological change, furthermore, occurred in a social milieu in which no one is allowed to be just plain tired or full of aches and pains. Until someone has taken them seriously, studied them, and developed good tests, many nonspecific ills, ranging from chronic fatigue syndrome and Lyme to generalized environmental allergies, have been stigmatized as hypochondria.

Diagnosis is a big problem for all zoonoses. Few have typically distinctive signs; the target-shaped rash of Lyme disease, which reflects infection with the organisms at the site of the tick bite, only happens in a quarter of the cases. A history of being bitten by a tick is helpful, but since the ticks are so small, most people don't remember that event, momentous as it is in the life of the tick. The laboratory tests can be unreliable; some of them are plagued with false positives (the test is positive, but the disease is not present), while others have false negatives (the test comes up negative even when the spirochete is touring the bloodstream).

TABLE 1

	DISEASE POSITIVE	DISEASE NEGATIVE	TOTAL
TEST POSITIVE	99	909	1,008
TEST NEGATIVE	1	89,991	98,992
TOTAL	100	90,900	100,000

Even the very best tests do badly if the disease is rare. It is worth doing a little technical aside here. Look carefully at Table 1. The sensitivity of a test is the proportion of those who actually *have* the disease who test positive; we ask, if you have the disease, will you test positive? The specificity is the proportion of those who *don't* have the disease who test negative. These characteristics are pretty stable.

They also, often, involve trade-offs. Let me take an extreme case: If I said that all men have prostate cancer (my test being sex classification), then my sensitivity would be 100 percent; that is, every person who has prostate cancer will be correctly classified as having prostate cancer. Unfortunately, I will also have falsely classified a whole lot of healthy men as cancer positive; that is, the test has a very low specificity. If I say that no men have prostate cancer, then every man who does *not* have prostate cancer will be correctly classified (specificity = 100 percent), but my sensitivity will be zero, since no one with the actual disease will be correctly classified.

Looking across the table at predictive values, we discover something different. A positive predictive value asks, if you test positive, how likely are you to have the disease? The answer to this question depends not only on the sensitivity of the test but also on how common it is, its prevalence. Let us suppose (as in Table 1) that Lyme disease occurs in one in every thousand people (100 out of every 100,000 people in the table), which, as diseases go, is pretty high. Let us say that we have a test that is 99 percent sensitive (it correctly picks up 99 percent of those who actually have the disease, but it misses 1 percent) and 99 percent specific (it correctly classifies 99 percent of those who are disease free, but it also throws 1 percent of the healthy people into the diseased category), which is better than almost any test on the market for any disease. If only one in a thousand people actually has Lyme disease, then less than 10 percent of those who test positive actually have it. If 90 percent of the people in the population have the disease (flip the "positive" and "negative" columns around), then about 90 percent of those who test positive actually have it.

Mull this over. You have the best test in the world and it is usually wrong on test positives; on the other hand, if the test says you don't have the disease, you probably don't. This is why tests for genetic screening in the general population are much more often wrong than right, and why every test we use, like that for SARS-COV-2, needs to be part of a judgment call that integrates clinical exams, history, lab tests, epidemiological reports, and the physician's experience.

It is important to know how common the disease is in the area before you test. If a disease is more common, then your test will perform better. But how do you know how common the disease is unless you test? This dilemma—that we depend on tests to describe the true state of the world, but we don't know how well the tests are performing unless we know the truth to begin with—is one of the big reasons that our scientific knowledge usually increases more slowly than the events we are studying; the COVID-19 pandemic represents a shift in speed, but the virus still spreads more quickly than our knowledge of it. In general, our understanding of a disease depends on comparing many different kinds of tests over time and a good knowledge of the biology of the species involved. If there is a large population of people with undiagnosed ailments, practicing physicians, under pressure to do something, are strongly tempted either to lump them all into the "diseased" category or to dismiss them as hypochondriacs.

There's no quick fix for this kind of problem. Since tests for diseases always perform better where the diseases are more common, progress depends on asking questions related to where the organism and its associated disease are more likely to be found. Are there social and ecological clues that will help us make the tests perform better?

Richard Ostfeld and his colleagues determined that Lyme disease was less likely to occur in more biologically diverse habitats, since the ticks and the bacteria they were carrying were less likely to be able to find suitable hosts on which they could feed in such habitats. Models used by C.S. "Buzz" Holling and his colleagues in the Resilience Alliance suggest that diverse habitats, which buffer against disease, are resilient, meaning that they have the ability to adapt and change and continually reorganize themselves in the context of a changing world. Members of the EcoHealth network might refer to it as ecosystem health, which means something similar. Whatever term is used to describe this ability to have long, healthy lives on this planet, we are losing it, not just through loss of biodiversity but also through human-induced climate changes.

So the disease is more likely to be found where the deer are plentiful and the field mice and the ticks can live out their full lives as expeditiously as possible—in warm places with as few species competing for food and housing as possible. Unfortunately, by most accounts, we are creating conditions in a lot of places where Lyme disease is more likely to occur. This development may be good for people who are worried about test performance, but it is bad news for those of us out for a walk in the woods.

Our own research team, led by the veterinary entomologist Nicholas Ogden, studied temperature and microhabitat effects on tick development and linked those effects to projected climate changes. We already knew that Lyme disease had a strong foothold in two southern parks of Canada—Long Point and Point Pelee—in part because those are major transit points for thousands of migrating birds every spring and fall and in part because the temperature, moisture, and general ecological conditions were right. We already knew that those areas had white-tailed deer and field mice, both necessary for the disease to establish itself. We also knew that migrating birds carried ticks (so-called adventitious ticks), which dropped off throughout their migration routes, like passengers getting off airplanes at various destinations.

Because of these adventitious ticks, we occasionally saw cases of Lyme disease scattered throughout Canada, well away from areas considered to be endemic. So far, we had little evidence that the ticks and the disease they carried were being established in local, home-grown cycles. The question we asked was: how important would milder winters and hotter summers be in the northward movement of Lyme disease?

To get the ticks, Ogden and fellow researcher Robbin Lindsay took four male research dogs out into the woods at Long Point, one of the southernmost areas of Canada. Robbin, an intrepid entomologist with the Public Health Agency of Canada's secure laboratory in Winnipeg, who had battled the prairie mosquito swarms, tramped through jungles in Africa, and spelunked in the dank sewers in Windsor, Ontario (where he found overwintering mosquitoes harboring West Nile virus), thought he had seen just about everything. The theory was that the four beagles would attract ticks, which would then be picked off and studied under controlled laboratory and field conditions. Unfortunately (for our study), two of the more dominant dogs took a keen sexual interest in the third dog, who was apparently giving off receptive messages. Nick and Robbin were left dragging an entangled threesome, plus the fourth dog, who seemed to be of the more traditional alpha male Christian Heritage family variety.

In the end, they managed to get enough ticks to run the experiments, although some of the ticks escaped the escape-proof containers we had designed for them. One of our research assistants, Matthew Waltner-Toews (and yes, related), was not surprised. He recounted tales of how, during monitoring of bird populations at Cape Cod, the tiny ticks would manage to slip down undetected inside the socks-over-pants, into the shoes, and back up the legs into dark and moist and unmentionable places. Our field experiments demonstrated that the ticks did very well in most environments, even our most severe ("control") sites, and that temperature seemed to be the major driving force for tick development. In part, what was protecting Canadians and other northerners from diseases like Lyme disease was a cold winter.

From our study (and others) it became clear that warmer weather associated with climate change was already encouraging the ticks and the disease to migrate northward into parts of Canada where they were formerly seen only rarely. If the climate and environment were stable, we could develop public health and management strategies to deal with diseases where they occur endemically. By altering the climate and the environment, we are changing the baseline and encouraging diseases to spread into new areas. In other words, our ability to adapt is being undermined by the way in which we continually change the environment to which we are trying to adapt. Driving cars is connected to global climate change, which is connected to local weather change, which is connected to the spread of disease in landscapes transformed through agricultural and economic policies and urban and peri-urban planning and landscape development activities. The possible strategies to deal with this problem are discussed in the final chapters of the book, since the success of responses to one disease problem is very much connected to how we solve a whole lot of other problems.

(6)

BITING FLIES, KISSING BUGS, AND SLEEPING SICKNESS

NOT ALL arthropods have cast their lots with the rodents. Some of them, the biting flies that carry parasites that cause sleeping sickness and kala-azar (leishmaniasis) among them, have gone upscale on the evolutionary ladder.

One of a family of diseases caused by a genus of slinky, willow-leaf-like blood parasites called trypanosomes, sleeping sickness is one of those African diseases that have perplexed and frightened people for many centuries. The parasites, members of a class of living things called Euglenozoa, are descended from some of the oldest living things with nuclei (eukaryotes). They are thought to be about 300 million years old, which takes us back to the Paleozoic era (roughly 543 to 248 million years ago), a world of swamps, explosions of a brilliant, exuberant diversity of all forms of animal life, and wandering supercontinents, when Africa and South America were still joined at the hip. The era ended catastrophically (at least in the unimaginably long lifetime of earth) with a breaking up and drifting apart of the continents and, one might suggest, an incontinent flood of species extinctions (as high as 90 percent). This early birth of the trypanosome species explains why there are different genuses, with different ecologies, but similar ancestries, in South America and Africa. I'll come back to the South American one a bit later.

Tsetse flies, which (in Africa) now transmit the parasites from one tropically warm bloodstream to another, go back about 200 million years, to Mesozoic times (some 248 to 65 million years ago). That era gave us ferns, cycads, ginkgophytes, bennettitaleans, and other unusual plants, as well as conifers and angiosperms (a curious name, with etymological scents of blood vessels [angio] and male reproduction). Some with a more prosaic turn of phrase would call them flowering plants, which provide us with pretty well all the foods and many of the drugs we rely on. The era also gave us Jurassic Park and the oil deposits of the North Sea. Sadly, it also ended with a mass extinction, which took down not only the dinosaurs but also the ammonites, who I once fondly dreamed might have been the forebears to Mennonites, who begat me. Alas, not so. According to some sources, these marine mollusks succumbed on the Maastrichtian stage in, one would imagine, a dramatic fashion. The stage is named after the site in the Netherlands where chalk rocks typical of this era are found and where the Treaty on European Union was staged and signed in 1992, although it would be stretching credibility to argue for some sort of evolutionary connection here.

So both the parasites and the flies are survivors. We humans are newcomers on the scene, a mere 5 million or so years ago, and probably just picked up the parasite when we started farming and herding cattle in risky landscapes. For thousands of years, people and parasites lived in a kind of loose truce, people tending to avoid tsetse fly habitats and clearing brush from around their houses. Our written records of the flies (probably) go back to the Jewish prophet Isaiah of the early to mid-700s BC, who wrote: "In that day, the Lord will whistle for flies from the distant streams of Egypt... They will come and settle in the steep ravines and in the crevices in the rocks, on all the thorn bushes and at all the water holes" (Is. 7:18–19, New International Version). The image of this Old Testament YHWH whistling for flies, evoking, for me at least, a picture of an old, white-haired field biologist surrounded by a cloud of small insects, is not the one I was offered as a child.

The historian Ibn Khaldoun in 1406 reported not just on the annoying flies but also on the disease itself. There were reports from Mali that a king suffered two years of profound sleep on the throne (did no one notice?) before he finally died of the disease. Khaldoun apparently picked up this information from traders crossing the Sahara. Certain tribes in western Africa apparently had about fifty names for the disease, which, if one might draw a parallel to the many names for "snow" northerners are said to have, or names for "sand" among desert dwellers, indicates either very close familiarity with the disease or a culture where people have much idle time on their hands. Our scientific knowledge of the parasite increased rapidly in the nineteenth and early twentieth centuries, when it began to interfere with the slave trade and hindered European pillaging of Africa.

As one might imagine, the trypanosome ancestors probably evolved, much as people have, from free-living swimmers to parasites of non-human animals (about 5 million years ago).

Some subspecies only affect wildlife and domestic cattle. The Zulus named the animal disease "nagana," meaning "a state of depressed spirits," which pretty well sums it up. Two members of the species *Trypanosoma brucei,* however, can be shared between other animals and people. *T. brucei gambiense* generally lives in western Africa, and the disease it causes unfolds slowly and inexorably, a bit, perhaps, like nagana. This species of the parasite is fully adapted to, living in, and spreading between people and has shown no interest in returning to the wild. Other species only infect animals, which are of interest to veterinarians and as historical artifacts but not as zoonoses. For those interested in genealogy, the earliest trypanosomes were probably adapted to wildlife, in which many of them, to this day, seem to cause few problems. The zoonotic form (*T. brucei rhodesiense*) emerged later, probably in the Ionian (also known as the Middle Pleistocene) epoch, tens to hundreds of thousands of years ago. Still later (say, ten thousand years ago), the human-adapted form, *T. brucei gambiense*, emerged.

One member of this family of blood parasites, particularly relevant to our understanding of disease emergence, ecology, and our stumbling attempts to define a convivial niche for ourselves on the planet, is *T. brucei rhodesiense*. This one lives mostly in eastern and central Africa and causes acute fever, aches and pains, and general miserableness. The body's immune system attacks the parasites, but these little guerrillas have complicated, multi- and mini-circles of DNA, including one thousand genes (which also mutate, creating more) coding for different surface coats. Just when the body thinks it has defeated them, they come back in different uniforms, and the battle starts again; the immune system is at first very active and later exhausted. Toward the end, the parasites slink their way across the blood-brain barrier into the king's castle and start messing with people's minds.

Classically, human sleeping sickness starts with a flu-like illness and proceeds through disrupted sleep patterns to the final sleep of an irreversible coma—hence the name—but psychoses and aggression are not unheard of, and patients have inadvertently ended up in prisons and mental hospitals. WHO has estimated that there are upward of 300,000 cases in people a year. In the region in eastern Uganda that I visited as part of a research team in 2001, about twelve thousand people had died since 1986.

This "Rhodesian" sleeping sickness can be carried around by a lot of different animals, including cattle, where it may stay cryptic or cause nagana. These trypanosomes are carried from animal to animal and to people by tsetse flies, which, like many sensible animals in the tropics, prefer to live in shady places, especially near waterways. The female fly, perhaps having been schooled by thrifty Scottish Presbyterians, breeds once and then, with almost obsessive thrift, saves the sperm and uses it as she needs it. I have no information about what must surely be the onanistic, lonely habits of the males. The female fly lays an egg every nine days in moist, loose ground, and the new flies emerge about three weeks later. This reproductive frugality flies in the face of the "reputed logic" of reckless child production systems, which have allowed humanity to fill and dominate the earth but are now threatening the whole wonderful biospheric experiment. The flies have been around for a lot longer than us, and we might reconsider our strategies.

Context, of course, is everything.

On May 28, 2000, the day before my fifty-second birthday, I was sitting on the deck at the home of Catherine Kenyatta, gregarious, gracious, cheerful granddaughter of Jomo Kenyatta, founder of modern Kenya, and her husband, Sean. We were in Mbale, eastern Uganda. A motley crew of researchers from Kenya, Canada, Zambia, England, Scotland, and Uganda, we were there to work with the Ugandans to try to understand, and control, the re-emergence of sleeping sickness in eastern Africa.

The view from where I nursed my beer was across the darkening shadow of the town to the foothills of Mount Elgon. I would say it was evening, but readers in temperate zones might confuse this with the slow draining of color that occurs at dusk. There, I could already sense the black pixels gathering under leaves, in the shadows of huts and palms. When they reached some critical point, the whole landscape would suddenly cascade into darkness. I wondered if this difference in the experience of diurnal change might explain the faith among many temperate zone scientists in gradually changing ecosystems (global warming degree by degree, increments of pollution, slowly rising chronic disease epidemics) and blinding them to the critical points, collapses, and sudden changes that characterize pretty well all the world's ecosystems and may give us clues as to the behavior of twenty-first-century pandemics.

From a distance, the nearest hill rose up in red-gray-black cliffs to a plateau, waterfalls spraying down crevices, shaggy green tatters like carpets stuck precariously to the cliff faces. But views from a distance should not be confused with "broader views" and may be misleading; sometimes distance gives not perspective but fuzzy thinking.

That afternoon, veterinary epidemiologist John McDermott, geographer Barry Smit, and I had clambered up one of those steep slopes. Up close, every fold and wrinkle of the hillside was inhabited by villages and farms, kids, goats, men and women washing carrots in a stream. The green patches on the cliff faces were farmers' fields. Barry, a Kiwi-cum-Canadian, aspiring sheep farmer, co-author of reports from the Intergovernmental Panel on Climate Change, does not suffer fools or idle activities gladly. Unlike me, he dislikes activities without a clear purpose. He therefore gave our walk a purpose, identifying an end (some caves high up on the slopes), and things to learn on the way (asking our guides all about the crops being grown).

I was at first annoyed at this intrusion into my non-doing, thoughtless, mind-clearing activity. I was soon captivated, however, as Barry asked again and again: What is this? Why are they growing that? In those scant, thin soils clinging to the rocks, farmers were growing: Irish potatoes (three kinds); yams (four kinds); sweet potatoes; cabbage; carrots; cassava; bananas (one kind for eating; another kind where the fruit was inedible but the sap from the stem was used for treating measles); plants to treat malaria and gonorrhea; peas; beans; pumpkins; leguminous trees; napier grass for erosion control, cattle feed, and mulching;

eucalyptus for fuel and soil control; onions; coffee; and at least two kinds of trees near houses to serve as lightning rods. I was in awe of this managed complexity and the possibility for hope for our species that this human-modified landscape embodied.

Members of the Resilience Alliance use a lazy-eight, or infinity-sign, model to interrogate and explore the world. In this model, linked social-ecological systems move through natural phases of growth, conservation, creative destruction, and reorganization. According to the Resilience Alliance researchers, the systems we live in and are part of can recover from destruction such as fires and disease outbreaks if that destruction is localized, embedded in a larger, resilient system. The whole complex model has been called a "panarchy." Through a variety of networks and feedback loops, the larger landscape within which the local areas are nested provides nutrients, genetic material, and energy for renewal. The larger system "learns" from the smaller outbreaks and in turn uses this learning to respond to other local outbreaks and fires elsewhere in the larger system. This learning is embodied not just in biological and genetic diversity but also in social and cultural diversity and in the links and communications that exist among these diverse elements. This idea of many small catastrophes embedded in a larger resilient set of systems is sometimes referred to as a shifting steady-state mosaic. Individuals die and families suffer, but the biosphere continues to make a home for our descendants. In a sense, the nature of which we are a part is a kind of flexible, responsive, diverse welfare state.

For farmers in the tropics, maintaining various diverse crops means that they can more easily adapt to changing market, social, and climate conditions around them. If some parts of the system crash (prices of certain crops, the crops themselves), there is enough redundancy of function in their collective agro-ecosystems that some of them will be okay most of the time, they can take care of each other, and, collectively, they can ride out the crash, learn from it, and keep going. Also, diverse ecosystems provide some protection against disease epidemics; if a variety of crops serving similar functions (providing food, healing, material for building) are grown, and a variety of animals, including those that eat insects, inhabit that agro-ecosystem, then bacteria, viruses, and parasites don't have the equivalent of one of those all-you-can-eat "trough" restaurants that are so beloved by many Americans and that monocultures freely provide.

Although diversity protects against the spread of epidemics and pandemics, however, it can, in some circumstances (when people invade new territories, for instance), foster the emergence of new troublemaking microbes, as the stories of avian influenza, SARS-COV, and SARS-COV-2 demonstrate. The challenge is to find ways to foster diversity-related resilience without promoting diversity-related disease. It is possible, but it will require us to go well beyond the ecologically simplistic thinking embedded in the either-or mindsets of "progressive, scientific economies of scale" versus "old-fashioned, subsistence poverty."

When we reached the end of our walk, up near the ridge of the mountain, we stepped into a cool, dank cave, once used by rebels during the time of Idi Amin. For a moment we stood there—John, Barry, the two guides, and I—bent over in the low dusk of the cave. Then, Barry's clear tenor voice echoed around us, singing the haunting Righteous Brothers song "Unchained Melody." We stood in silence, as if this were some kind of religious ritual, which, I supposed later, it was. Then we walked back down the mountain.

That night, I lay on my back and pondered the options available to the Ugandans—available to *us*, if I am to count myself as a fellow human being—for dealing with zoonotic sleeping sickness. Barry Smit, myself, and the others were here because John McDermott had put together an international research team to answer the question: "Links between sleeping sickness and natural resource endowments and use: what can communities do?" What indeed? I thought.

We could wait until people were sick and then treat them. But finding sick people early and treating them aggressively with toxic drugs is not a program with a high chance of success in central and eastern Africa. For one thing, this was not the plague, and the drug of choice was not some benign antibiotic, as it still is, in many places, for *Yersinia pestis*. Some suggested that the treatment for sleeping sickness felt like having an infusion of red-hot chili peppers. Médecins sans Frontières had enough trouble getting pharmaceutical companies to produce and provide the drugs and then putting them to use, against the much simpler problem of human-adapted *gambiense* sleeping sickness.

We could try to get rid of all the animals that are carrying the parasite. This had been tried before, both by people (as in game destruction) and by the unpredictable gods of history. In 1896, an epidemic of rinderpest, caused by a virus that is probably the ancestor of measles in people and distemper in dogs, swept through southern and eastern Africa, killing millions of cattle and big game. The parasites, in a small-is-beautiful move that presaged the economic philosophy of E.F. Schumacher, survived in the small, numerous, adaptable bushpigs and bushbucks, as well as spending more time among people.

We could spray massive amounts of pesticides to kill the flies or cut down all the shady places where the flies like to breed, but those strategies had already been tried. In the 1950s, before there were subjects like biology in school, both knapsack sprayers and airplanes were used. And clearing all the brush along streams is also a non-starter. Poor communities who live near rivers and depend on cattle and wildlife for food are not usually keen on destroying the places where they live in order to get rid of a disease.

The big outbreaks of sleeping sickness in the twentieth century happened during periods of social upheaval—the early and late years of Idi Amin's terrible reign, for instance. In these times there was breakdown of health and agricultural infrastructure, along with widespread violence. During the fighting, people fled their homes; after the fighting, they returned, bringing with them cattle from southern, infected territories. When raiders from Sudan in the north rustled those cattle away, the people brought in more infected cattle.

Now, in the post-Amin years, Eric Fèvre, Lea Berrang Ford, and other members of our research team found that new areas were becoming infected as cattle were moved into them from the south. After entering a new area, the parasites and tsetse flies spread out along waterways and other fly-friendly habitats. Our big concern was that the *rhodesiense* parasite was moving into territory normally inhabited by the human-adapted *gambiense*, with complicated consequences for diagnoses and control. Could the epidemic be stopped? The diversity of the farms we had seen on the hillside reflected a newfound optimism among Ugandans, which was also expressed in a vibrant local democracy. This optimism, coupled with civic engagement, offered some hope for finally coming to terms with the old parasite.

Still, the disease presented itself as what is sometimes called a "smoldering" epidemic: the infection spread inexorably through the countryside, but clinical cases were always at a level low enough to stay just under the radar. On the one hand, a low number of cases seemed to offer an opportunity to eradicate the disease, at least from certain areas. On the other hand, for people resettling land and struggling to get past decades of war and destruction, a few cases here and there hardly provided motivation to mount major control programs. The old top-down approaches of technocrats telling locals what to do didn't have much support; the people at the top didn't have the money to sustain such a program, and the people on the ground had other things on their minds, and had had enough of others telling them what to do.

The day after our trek up the mountain, we piled into our vehicles and headed out past open fields north toward Soroti, catching glimpses of Lake Kyoga, and over rutted, dusty roads where weather-beaten villagers huddled next to burlap sacks of charcoal, to a group of mud and thatch huts gathered in the shade of a grove of trees. This was to be our medical research post for the day. We set up tables with centrifuges and blood collection tubes. Some of us interviewed farmers under a tree. Men from the village, joined by researchers, wrestled with the cows, checking ears and testicles for ticks (lots of them), taking tubes of blood from the jugular and sucking up a tiny bit from the ear into a glass capillary tube.

Women, men, children lined up under the trees, forming a winding line around the village, all to be questioned, examined by a nurse or doctor, and have blood taken. Cattle milled about in the shade of various trees in the vicinity. Children cried, cattle in heat clambered on to each other and tried to breed. One young cow chased a dog around and around, trying with its nose and tongue to determine the nature of this shy creature. Children stood in a circle around me as I did a second reading on hemoglobin color tests for the medical technologist. They were most fascinated when she injected blood from cows, suspected of being infected, into mice, where it could be kept alive for later molecular studies. The mice, held by the loose skin on their necks, looked open-mouthed with surprise.

Near lunch, in the heat of the day, I went out with three other fellows to hang traps for tsetse flies. We wandered down a sunbeaten path to the swampy area near the river, one of the spidery fingers of water draining into Lake Kyoga. The soft ground was crisscrossed with deep ruts, paths from people and animals looking for water. I wished I had my Tilley hat, and I carried the traps over my head for protection from the blistering sun. The traps are like tents, with white mesh at the peak and blue and black cloth at the base, since tsetse flies are attracted to black and dark shades of blue. Thus, there was some poignancy is seeing schoolkids wearing blue and black shorts and tops clearly made from fly-attracting traps, which underlined the urgency of engaging local populations in the enterprise, combining education with action.

When the flies reach the colored lower part of the trap, they fly up toward the light, through small holes, and into the top part, where they are trapped. We hung the traps from trees in shady areas, often near water. I didn't pay much attention to the first water hole we came to, thinking it to be a dugout for cattle.

At the second or third dugout we visited, there were several women, bent over, letting the brown liquid burble into their bright canary-yellow jerry cans. I asked about the coffee-latte-colored water in the hole. Was this for drinking?

"Yes," said the ten-year-old kid who was showing us around.

"Without boiling?"

He cleared away the flotsam and insects at the surface, scooped some water into his hand, and drank it. "But not for you," he grinned. "You will get sick."

The two other men with us, who were from outside the area, also said they would be sick if they drank that water. I thought of the people in Walkerton, Ontario, where thousands had been sickened in May 2000 by drinking "clear" tap water; several died. Pausing beside a termite mound six feet or more high, the boy grinned at me as he dug away with a stick, pointing out the big, angry termite guards that came out to challenge his intrusions. When fried, he assured me, they were excellent with sesame paste.

This is why I am here, I mused. A sense not so much of home, or of saving the world (although we would all like to do that, I suppose), but of camaraderie, of being on an incredible journey together, of being part of something wonderful, bigger than all of us, of trying to make a not-too-uncomfortable place for ourselves in the midst of the messy, complex, amazing world we live in. Of promoting health, defined by the late, great microbiologist René Dubos as "modus vivendi enabling imperfect [people] to achieve a rewarding and not too painful existence while they cope with an imperfect world."

Returning from hanging the traps, we found that the others had already eaten lunch. I stood under a tree, wiping the sweat from my forehead, and drank a Coke. I recalled the Coca-Cola sign at the equator, when we had driven over from Kenya the previous week. I had thought then, and I thought now, that one of my motivations for being here was to provide a counterweight, to help people find an alternative story, to that of the ubiquitous Coca-Cola-type McEmpires. But out there in the sun, damn, was I thirsty.

On the evening of May 29, 2000, at the Sunrise Restaurant in Mbale, I looked around the table at our sunburned, disheveled group and reflected once more on the web of narratives and motives that had drawn us there. We looked over the menu, and someone asked what a White Russian was. I did not, then, recount my entire family history, the revolution, the civil war, the fleeing of wealthy farmers from Stalin's bureaucratic utopia, that ultimate in top-down scientific technocratic solutions. But I did think then that my own personal story was part of this larger story, that every researcher is part of what he or she does research on, that sustainable health and development are all about finding a common journey and a Chaucerian story we can tell each other on the way. What passed my lips was something like, "Did you know that my parents fled from the Soviet Union in 1926? And now here I am in Uganda, on my birthday."

The British epidemiologist Paul Coleman's response to this statement was to order a round of White Russians, with the rule that I would have to finish the drinks of all those who didn't. As a benediction, somewhat inebriated, fresh from a day in the field and beers at Catherine and Sean's, emotional (I hesitate to call it maudlin, but perhaps it was), scrounging through my topsy-turvy mind for a succinct summing-up of this out-of-the-lab-and-into-the-world science, my mental and physical transports from Canada to Mbale, I could think of no better response to the complex history of sleeping sickness, and our challenge to come to some convivial, ecologically negotiated truce with it, than to recite the poem "Wild Geese," by Mary Oliver, a wonderful evocation of wild geese flying overhead and their "harsh and exciting" message to us that we are all part of the same family.

TRYPANOSOMES IN THE Western Hemisphere took a somewhat different path to finding a home. Chagas disease, named after Carlos Chagas, one of the world's great biomedical investigators, is a kind of distant cousin to African sleeping sickness. Caused by the parasite *Trypanosoma cruzi*, the disease is sometimes called American sleeping sickness, but that is a misnomer, since the disease it causes is nothing like African sleeping sickness. Technically, one could simply call it American trypanosomiasis. As you can imagine, after a few hundred million years of separation, the family resemblance has been somewhat diluted. Unlike its African cousins, this parasite does not cross the blood-brain barrier, so the disease has none of the manifestations associated with infection of the brain. Many people get infected, but not many become ill. In those who do succumb, the heart grows large and flabby, and it can't effectively pump blood anymore. In another form of the disease, the intestines get large and flabby and food pools in the gut. In either case, the victim dies, slowly, painfully, over many years.

One Brazilian researcher, the late Philip Davis Marsden, described the disease as "retribution for colonizing the New World." From Mexico south to Chile and Argentina, some

8 million people are thought to be infected. The CDC estimates that there are about 300,000 people infected with *T. cruzi* in the United States. Evidence from dried-out human mummies suggests that the parasite was already cycling among animals and bugs in southern Peru and northern Chile nine thousand years ago, waiting for more humans to come intruding. They did.

Although they behave differently, *T. cruzi* don't look much different from other trypanosomes once they are in the human bloodstream. Swimming along like delicate tropical fish, the parasites undulate their way into various body cells, where they lose their tails and, like many new human home-owners, wreak their damage locally. Actual clinical disease associated with these parasites, though not common, may be acute or chronic and involve a wide range of organs, from the heart to the brain to the intestines. Until research-based control programs were instituted in the late twentieth and early twenty-first century, it was an important cause of sudden death from cardiomyopathy (pathology of the heart muscles) throughout Latin America, as well as causing a chronic ballooning of the intestines called megacolon.

From a medical-ecological point of view, there are various cycles of transmission, some restricted to non-human mammals (over a hundred species infected), some crossing into human populations, and some passing from person to person. The traditional means of transport for the trypanosomes is through what have been called "kissing bugs" of the triatomine family. These blood-feeding insects might be tens of millions of years old, but they have adapted well to human invasion of their territories. At night, they quietly emerge from the cracks and crevices where they live and "kiss" people, soundlessly, painlessly, looking for a blood meal, at the corners of their victims' eyes or in small wounds or scratches. The bugs' eyes are bigger than their stomachs, as my mother would have said, and they have to urinate a lot just to keep sucking in the blood. These bugs, not being hygienically fastidious, defecate where they eat. The parasite lives in the triatomine's feces. Rubbing their eyes when they feel the tickle of the bug, people also rub the parasite into their blood. More than fifty species of triatomines can carry the parasite.

From a broader cultural point of view, this is a disease of global politics and power struggles between empires, from the Europeans and Incas several hundred years ago to the United States and Latin America today. Triatomine bugs originally lived (happily?) in free-living forest mammals in South and Central America. With deforestation, some bugs that were originally sylvatic ("wild," as in Sylvester the cartoon cat) seem to have adapted to more open types of vegetation and have also developed a penchant for certain kinds of human dwellings, thus increasing the frequency of transmission of *T. cruzi* to humans.

These types of poorly constructed (for people, not bugs) dwellings, frequently found crowded into shantytowns, did not occur by mere chance. During the height of the cold war, the United States and the former Soviet Union fought their battles in many poor countries. While the Soviets were subjugating parts of Europe, in Latin America, the U.S. overthrew democratic governments and fought pitched and dirty battles against popular uprisings. These wars drove many people from the countryside to slums around major cities. In recent years, the move from the countryside to the city has been accelerated by neoliberal economic policies, which have tended to favor large, wealthy landowners and to drive peasants into forest areas (and the *T. cruzi* sylvatic cycles) or into the slums (the *T. cruzi* urban cycles). The slums that have resulted from this combination of military and economic battles are an ideal breeding ground for the triatomine bugs that carry the parasites.

Now, in a new twist on the retribution theme and another warning of how small organisms can adapt more quickly to cultural change than the people who initiated the changes, the parasites recently discovered a new superhighway created by people: blood-bank systems in Latin America in the 1980s were infested with the parasites at rates from 6 percent (Buenos Aires) to 63 percent (Santa Cruz, Bolivia). These blood banks are likely to provide a pathway for the disease to spread from the poor to the wealthy. If they had the brains to anticipate, the parasites would no doubt look forward to a globalized blood supply.

Going to an even larger picture, it seems that changes in climate may have already begun to increase the distribution of one of the insect vectors of Chagas, *T. nigromaculata,* which has been found for the first time at altitudes above 2,600 feet in Venezuela. Infected vectors were first found at 2,600 feet in 1939, but an uninfected colony was found more recently in a modern dwelling at an altitude of more than 3,600 feet in the Venezuelan Andes. So fossil fuel use, which contributes to global warming, can also be considered a cause of the spread of Chagas disease.

In some ways, if the conundrums of diseases such as these can be solved, there is hope that many other difficult global problems can be resolved. Indeed, we may *need* to resolve just about everything in order to solve anything in particular. This is both a good thing and a bad thing. It means that the solutions to disease and food and energy and community may all be part of the same, much larger, more complex solution. If only we can find them.

THE PARASITES THAT cause American and African sleeping sickness are not the only troublemakers in this family of parasites.

From the mid-1980s to the mid-1990s, a couple of hundred thousand people in the southern Sudanese state of Western Upper Nile (also called Unity State by the government in Khartoum) died of what was described as an "AIDS-like" disease. The disease, kala-azar, or visceral leishmaniasis, was being carried by sand flies in acacia forests along flooded rivers. Investigators think soldiers took the infection into an area that had been reforested after destructive floods in the 1960s.

The leishmania are one-celled parasites from the same family as the trypanosomes and thus go back hundreds of millions of years. According to one (well-grounded) story, at least one of the major forms of the parasite has its ancestral home in what is now southern Sudan and spread out with early human migrations into the wider world of central Asia and southern Europe. It was probably living in our faithful hangers-on, domestic dogs.

In the early nineteenth century, there may have been a second African exodus, some parasite-infected animals hitchhiking on the slave trade boats to head east across the Red Sea and into India. A British army physician, William Leishman, discovered the parasite associated with Dum-Dum fever at a military camp at Dum-Dum, outside Calcutta (and where the airport is today). Later, as the creation of tea plantations allowed the disease to spread into Assam State, it was called British government disease. But other forms of the parasite emerged elsewhere. In eighteenth-century Turkey, it was called Aleppo boil. Among the Incas, it (or something like it) was called valley sickness and, paradoxically, Andean sickness. There is evidence that the disease was present in the Americas as far back as the first century AD.

Unlike true (African) sleeping sickness, which is found only in Africa, and Chagas disease, which occurs in the Western Hemisphere, leishmaniasis is a global citizen. Leishmaniasis is one of the world's top ten most important infectious diseases (along with both African and American trypanosomiasis). WHO's Tropical Disease Research Programme reports that leishmaniasis is endemic in almost a hundred countries and territories, including sixteen European nations. There are probably more than a million people infected every year. WHO estimates that almost sixty thousand people die from it every year and that it costs the world more than 2 million disability-adjusted life years annually.

Until World War II and the introduction of DDT, serious infections in children were common in Mediterranean Europe. The drop in numbers of people affected was attributed to the use of pesticides, but the infection remained in dogs, which means the parasite was still around and reproducing. More likely, the disease in kids disappeared because of better nutrition and improved immunity, the same things that staved off a lot of other diseases (without necessarily stopping infection). HIV, which allows latent infections to become active for many diseases, gave leishmaniasis, like tuberculosis, a new lease on life.

These little blood parasites live quite happily, not causing any major problems, in a variety of marsupials, two- and three-toed sloths, armadillos, and forest rodents (in Central and South America), rock hyraxes and African grass rats (in Africa), fat sand rats (in Saudi Arabia), and great gerbils (in Russia, Mongolia, and central Asia). In many countries, domestic dogs carry it, but they get sick enough to suggest that they aren't the natural host animal. In 2000, foxhounds in eighteen U.S. states and two Canadian provinces were sick with leishmaniasis, but that strain appeared to be transferred only between dogs directly, and no human cases were reported. In some parts of India and Africa, there appears to be one form that is adapted just to people, but that could be because we haven't looked at the right animal host yet.

Usually—the epidemic in foxhounds notwithstanding—blood-sucking sand flies (of which there are about seventy important disease-transmitting species) pick up the parasites from one animal and carry them to another. People often get the parasite, via the sand flies, from dogs, after the dogs get it from nosing around in the woods. The types of sand flies and the habitats they prefer vary from place to place. The flies seem to prefer dry areas in the Old World and tropical forests and savannas in the New. In some areas, they prefer large cities with slums and lots of poor people.

A couple of weeks after the female sand fly lays her five or six dozen eggs, they hatch, and the larvae nose around to feed on any kind of organic stuff they can find. They pupate and emerge as adults in darkness, just before the birds' morning song. Indeed, the wiggling larvae may just be the reason why the birds are singing. Male flies encourage females to mate by flapping their wings and giving off odors (what else is new?); both feed on sugary secretions of plants, but only the females suck blood. When not breeding or sucking on things, the flies rest in cool, shady, humid places. If the females suck blood from an infected animal, they get infected.

In sand flies, the parasites are slinky things, each with a flagellum (tail), which allows them to move around and hang on to the fly's gut cells. The parasites need to hang on tightly until about the third day after a blood meal, which is when the fly defecates. Once the fly has had a good dump, the parasites let go, relax, and multiply as fast as they can. They reproduce by dividing down the middle, which seems considerably less pleasurable, but more efficient, than some other means. The newly "born" parasites then swim around anxiously waiting for the fly to take another meal; when that happens, they are off to the big blood meal in the sky: a mammal.

In animals, including people, the parasite invades white blood cells of the immune system, where it loses its tail, rounds itself into a ball and multiplies. Once in an animal, it can do a variety of things, depending on the immune state of the animal, the particular species of *Leishmania*, and a lot of things that aren't yet very well understood. Much of the time, it appears that nothing particularly bad happens.

In other cases, very bad things indeed happen. Sometimes, if the parasite stays on the skin, it causes a reddish, insect-bite-like sore, which can expand into an open, wet sore called an Oriental sore, which looks a bit like a volcano. In some people, especially with one species of parasite in Brazil, the mucous membranes of the mouth or nose may be completely eaten away, sometimes years after the initial invasion. This condition is called espundia. The worst form of the disease is visceral leishmaniasis, in which the little stinkers invade the spleen and liver and cause a wasting-away-until-death-do-us-part scenario.

How do we begin to get a handle on these complicated parasitic diseases? There are the old diagnosis-treatment-vaccination tricks, of course. They are handy, but they depend on good diagnostic laboratories, vaccines, and drugs—in other words, on money and trained people. These are in short supply in those parts of the world where these parasitic diseases are most common.

In those places where dogs are a major source of the parasites for flies and people, they may offer some opportunities for control. In the late 1980s, one of my graduate students went to investigate an epidemic of leishmaniasis in a town in Central America. She wanted to explore the role of dogs. When she arrived, she found that local authorities had already killed all the dogs, and new ones were being brought in. The Chinese apparently were effective in getting rid of the disease in some places by getting rid of dogs. Dogs, however, are not just carriers of disease or even just human companions. In some countries, they are food. And, as we discovered in our work in Nepal, they may have other jobs, for example as night guards and community police. Even removing rodents or rodent habitats may pose unexpected problems, such as those we encountered with the plague in Africa.

How, then, can these diseases be prevented? The answers go beyond the community engagement and setting of fly traps that might have a chance for controlling sleeping sickness in Uganda. The biomedical approach would include treating sick people, spraying for bugs, testing people at risk, and screening blood supplies. This might help control the disease; it would also bring large benefits to certain chemical and technical industries in wealthy countries. But any sustainable health program to prevent these diseases would surely quickly become a social activist program. An ecosystemic and socially holistic approach to health will not, contrary to expectations, always create win-win situations. We still have to make moral choices.

Any programs to prevent these diseases caused by blood parasites must include a political agenda to create more egalitarian societies and a more just global distribution of wealth. With better nutrition, fewer people get sick, even if they are infected. With better-built houses, biting flies and bugs can be more easily be kept outside. If people do get sick, good medical care can stop the cycle of reinfection more quickly. It is just such a combination of medical and socio-political programs that seems to be bringing down the incidence of Chagas disease in the southern cone countries (Argentina, Bolivia, Brazil, Chile, Paraguay, and Uruguay).

It remains to be seen whether similar programs can be mounted for leishmaniasis and sleeping sickness, which have a more complex ecology. Given the general ecology of these diseases, we are unlikely ever to be rid of them completely, and the need for supportive medical programs to deal with inevitable tragedies in an overcrowded world is not going to disappear. We are left with a need for both ecological awareness and global solidarity.

(7)

WESTERN EQUINE ENCEPHALITIS, TELEVISION, AND AIR CONDITIONING

IN 1967, U.S. surgeon general William Stewart declared that "because [they] have been largely controlled in the United States, we can now close the book on infectious diseases." He was not the first person to confuse his country with the world, or his state of knowledge with knowledge in general, nor will he be the last.

Stewart's declaration of victory over infectious diseases, from the helm of the smallpox eradication aircraft carrier, was premature. The world seemed so much simpler then. We had the microscopes to find the problem-causing microbes, and the big guns—the antibiotics, the vaccines, the disinfectants—to kill the microbes or at least prevent them from causing disease. Smallpox was retreating before our superior weapons, and diseases such as polio and measles were lined up in the eradication queue. Other scourges such as leprosy and tuberculosis seemed to have just fled large parts of the world, leaving few explanations behind. The best guesses for what got rid of them are better food, cleaner water, and good housing—nothing really spectacular, nothing to patent.

Globally, infectious diseases really did seem to have taken a holiday; most people were more likely to die from accidents, suicides, heart disease, or strokes than from infections. But the microbes were still there, as anybody not living in suburban enclaves in North America, Europe, and Australia could have told us. In a 1990 study of causes of death worldwide, heart disease and strokes were number one and number two, respectively, but right behind them were respiratory infections and diarrhea.

The more scientists look, and the better the detection technology, the more microbes there seem to be. In the late seventeenth century, Anton van Leeuwenhoek saw what he called "animalcules," tiny animals, through his microscopes. Now there are electron microscopes, which can take pictures not only of bacteria but also of viruses and even those impossible disease makers, the prions. Viruses are thin bits of DNA (deoxyribonucleic acid) or RNA (ribonucleic acid), covered with distinctive jackets of protein. They move long distances quickly. They are the ultimate cool, athletic anorexics. There are uncountable numbers of viruses jostling around. They get into a bloodstream through a nostril or a gut or by being mainstreamed through the good graces of a blood-feeding arthropod. Once inside, they take over the host cells and use the organism's machinery to reproduce themselves. Mostly, they shuffle and sneeze and poke their way around pretty quietly, bird to mosquito to bird to mosquito to mouse. Viruses are probably as old as the oldest ancestors of the animals they infect, so there has been a long time to co-evolve and learn how to be good neighbors. Usually no one gets hurt.

In the southern United States, west of the Mississippi, there is a virus called western equine encephalitis (WEE). The virus multiplies in the blood of a variety of birds and in certain mosquitoes (*Culex tarsalis*). The original homeland of the virus was in a community of marsh-dwelling mosquitoes and birds such as blackbirds and swallows. It lived there for a long time without causing any serious problems in the animals it infected. As people moved in and, through drainage, irrigation, and land cultivation, destroyed some of that habitat and modified what was left, the viruses and the birds adapted. The birds found new sources of food, such as grain in fields or in grain bins, and the viruses found new hosts with cells to hijack.

On the one hand, the viruses needed some warm-blooded species with fast population turnover, so that there was always a fresh, non-immune crop to invade; house sparrows, which emigrated from Europe in the mid-nineteenth century, seem to fill that role in many areas. On the other hand, the viruses also needed mosquitoes willing to feed on both birds and mammals. With more than 150 species of mosquitoes in the United States (several thousand in the world), feeding at different times of day on a variety of species, everywhere from deep mine shafts to high mountains, the odds were pretty good that appropriate mosquito carriers could be found. The life cycle expanded so that WEE virus could now infect wild rodents, chickens, and pigs without, apparently, making them ill.

Not all animals were as lucky as pigs and chickens. In the summer of 1930, about six thousand horses in the San Joaquin Valley, in California, came down with a mysterious brain disease. They became uncoordinated, somnolent, and fatigued. Half of the horses died. The disease was WEE. From the 1930s to the 1970s, outbreaks and epidemics of WEE occurred in people and animals all through the western United States and north into Manitoba, Saskatchewan, and Alberta, in Canada; thousands of horses and hundreds of people got sick. Although birds and their attendant mosquitoes can maintain the virus in nature, mammals are apparently a so-called dead-end host: they can get the disease but usually don't get enough viruses in the blood to pass on.

From my memory of growing up in Winnipeg, Manitoba, hot, muggy weather, migratory birds, and great veils of mosquitoes drifting across the back lawn are staple features of the prairies. The question is not so much why the disease occurred, but why it didn't occur every year, and why it only attacked in some places and not others, when the birds and damned mosquitoes were everywhere. Could any of the outbreaks be predicted?

In the first instance, investigators used the fact that chickens get infected, and hence develop antibodies to WEE, without getting sick. For many years, putting cages of chickens outside in strategic locations and periodically collecting involuntary blood donations from them to look for antibodies provided a reasonable way to anticipate a coming epidemic. Rainfall and temperature conditions could be used to project mosquito populations; as the number of mosquitoes, and, if the virus was present, the rate of infection in chickens, increased, the probability of a serious epidemic could also be anticipated. This kind of epidemic prediction required a fair amount of intensive monitoring and still left unanswered the big questions: where did the viruses come from before they got to the chickens, and could the disease be stopped or predicted earlier?

Did any of the patterns of this disease have to do with where the virus stayed in the winter? Where *did* it stay in the winter? When I was in veterinary college, our instructor, John Iversen, had us take Richardson's ground squirrels out of the walk-in cooler and watched them slowly wake up. Iversen, a scientist who could suddenly switch his costume and, it seemed, his personality, from a three-piece suit to a white laboratory coat to rugged field gear, had a theory that the virus overwintered in these abundant rodents. After all, they do harbor a whole lot of other microbes. In the spring, when the last dirty patches of snow disappeared and chunks of floating ice clogged up canals and creeks, mosquitoes came alive in puddles by the road and migrating birds arrived from the south, initiating a cycle of amplification. By the time lovers strolled bare-armed through the woods, leading their horses behind them, the mosquitoes were singing in the sweet prairie air, and the epidemics could start in earnest.

Other scientists suggested that the virus overwintered in frogs or snakes or that it just lived through the winter in the gut of the hibernating mosquitoes (the way West Nile virus does).

At the time WEE was being investigated, researchers were asking the same questions about a similar, less common, but more serious disease—eastern equine encephalitis (EEE). This one, sickening a few horses a year at most, occurs sporadically from Louisiana north into Michigan, Ontario, Connecticut, and Quebec. It also appears to kill ring-necked pheasants and a few other birds as well as the occasional horse or person, but it is otherwise a little beast similar to the WEE virus.

One of the most plausible and intriguing theories is that these viruses don't overwinter in the north at all. Like many weary northerners, they spend the winter quietly reproducing in warm southern areas, along the Gulf Coast for WEE and Florida for EEE.

In the 1980s, two Canadian researchers, Robert Sellers and Abdel Maarouf, published a series of papers in which they asked: between 1980 and 1983, did WEE-infected mosquitoes ride the wind currents up from the Gulf of Mexico into Manitoba, Minnesota, and North Dakota? They found that a northward movement of warm air, meeting a cold front with rain, could well have carried infected mosquitoes and then dropped them at various locales in stages as summer crept its way farther north. It is a difficult theory to prove absolutely, given that mosquitoes are hard to microchip and that unstable weather patterns, vaccination of horses, changing landscapes, and changing habits of people all contribute to transforming a probability into an event.

Still, I am reminded of the time I was working on developing an animal-disease-monitoring system in the Caribbean in the late 1980s. One of my colleagues called me into his office one day to show me a giant locust in a jar of alcohol on his desk. "It's an African locust," he said. "I picked it up on the beach, right here in Trinidad." He paused. "Thank God they don't have enough space to swarm here." One of the Trinidadian women laughed. "Typical," she said in her slow lilt. "The big males fly over and then just lie on the beach all day."

It occurred to me to wonder: if those big lugs can blow over from western Africa and flop down on the beach in Trinidad, how hard can it be for a mosquito infected with WEE to get from, say, Texas to Manitoba? This also occurred to me later, when West Nile virus exploded into North American consciousness. A lot of other individual events have to fall into a particular pattern for an epidemic like that of West Nile virus to occur, but with the ecological, climatic, and social world in constant, rapid flux, and nobody really paying very close attention, it was bound to happen sooner or later.

The long-term patterns of WEE, like those of so many other mosquito-borne viruses, are puzzling. Although St. Louis and eastern equine encephalitis both make their periodic, sporadic appearances, the once-feared widespread epidemics of western equine encephalitis have all but disappeared. The CDC has suggested that the intensity of the natural cycle for WEE increases every five to ten years, leading to a higher probability of epidemics, yet the last epidemic in horses or people was in 1987. But the lack of disease in horses shouldn't be a surprise.

We take care of our horses. There's money in doing so. Effective vaccines against equine encephalitis were developed quickly, advertised widely, and rapidly made available to veterinarians. There are no similar vaccines for people. Who would pay? Who bets on people running around a dirt track? Vaccines are good for the horses (and the vets and the people who make vaccines), but the antibodies produced by the vaccine often interfere with our ability to differentiate animals that got the real infection from those that were vaccinated. Furthermore, if horses are vaccinated, there is no use monitoring them for infection or disease, so where do we look to find out if the virus is still out there, looking for another opportunity? We can monitor the sentinel chickens and look for antibodies in them, and we can search for the virus in mosquitoes. But if there haven't been any cases of serious disease reported recently, there is less pressure on government agencies to maintain monitoring programs. Money gets directed to other priorities until the next outbreak occurs.

People tend to get sloppy about vaccination, or cavalierly oppose it, especially when the vaccine works and the disease disappears. It is the general problem of preventive health programs, to which I referred earlier, that if they are successful, nothing happens, so governments and individuals are always tempted to disinvest in the most successful of them. Why would a horse owner pay someone a lot of money to stick a needle into her horse if there's no disease around? In 1987, there was a northward wave of WEE from the Rio Grande; starting in April and continuing into June, the virus moved north through the Texas Panhandle, New Mexico, and Oklahoma; in July, it headed farther north across the Great Plains; by August, it was in North Dakota, Manitoba, and Minnesota. Thirty-seven people and 132 horses were reported sick. In the Imperial and Sacramento valleys of California, the sentinel chickens developed antibodies to the virus, demonstrating that it was in the wild mosquito and bird populations, but no horses or people got sick. The sentinel chickens were the scientific equivalent of omens, but what was the message? What should be done?

Some have suggested that not much WEE is seen anymore not just because people vaccinate their horses, but also because there have been aggressive programs for spraying against mosquitoes. Others have proposed that this disease, like many others, has succumbed to broader social changes. One study in California demonstrated that between 1945 and 1982, counties that had higher ownership of televisions and air conditioners were less likely to have cases of equine encephalitis than places where fewer people had these newfangled luxuries. Apparently folks were now inside at dusk, watching television sitcoms in air-conditioned comfort and arguing about what's for supper instead of playing ball with their kids or sitting on the porch reading a newspaper. Their horses, of course, were vaccinated. And nobody really cared about what's out there, cycling between birds and mosquitoes. But with increased air conditioning and more TVs comes increased energy use, which translates into longer-term climate change. I doubt that those birds, their viruses, and mosquitoes are done with us yet.

(8)

WEST NILE VIRUS AND THE ST. LOUIS BLUES

WE HAD beer and chips at Jinja, where the Nile River flows out of Lake Victoria. We read the broken, rust-colored plaque there, amid the greenery near the steady flow of water where Ripon Falls used to be, before the Owen Falls Dam was built just downstream. The plaque proclaimed John Hanning Speke the discoverer of that most elusive and mysterious prize, the origin of the Nile. Still, I felt more pleasant tropical turpitude than historical grandeur. It was late May 2000. The place was quiet, just a few of the once-numerous Indian-Ugandan merchants and Western tourists starting to come back after the brutal ethnic cleansing of the Amin years.

A breeze wafted over us from the lake as a gray-headed kingfisher, with brilliant reddish-orange beak and strong blue and black wing markings, dropped down toward the water.

It was a place steeped in lovely natural vistas and pompous European views. The Arabs, looking for gold, ivory, and slaves, had produced the so-called Al Idrisi map of "Lake Victoria" (which was called Ukerewe until renamed in 1858) in the 1160s and described the lake as the source of the Nile. Ptolemy, that brilliant Alexandrian scholar, had more vague suggestions of lake sources of the Nile almost a thousand years before that. In the history of Western civilization, however, the lake's true identity and the name and the origin of the Nile could not really be known until a European had named them appropriately.

The lake was renamed after Queen Victoria, and the name stuck. Wealthy and inexperienced British explorer John Speke, who was on an expedition with the legendary Richard Burton, upstaged his companion (while Burton was recovering from illness) and proclaimed Lake Victoria the source of the Nile in 1858. Burton later challenged Speke to a debate over the true origins of the Nile before the Royal Geographical Society. Speke died just before the event in a mysterious hunting accident. A German, Dr. Burkhart Waldecker, trumped both of them by declaring in 1937 (from a safe temporal distance, since Burton and Speke were both dead by then) that the real source was a bubbling mountain spring in Burundi, or perhaps the Mountains of the Moon in Rwanda, fed by melting snows. Starting as snowmelt and spring water, the streams and creeks gather into small rivers, which eventually join to become the Nile. As climate change "disappears" the snow at the headwaters, one wonders what will become of the Nile.

North from where we sat on that day in May, the Victoria Nile tumbled and flowed through lakes Kyoga and Albert and north into the slow, shimmering spread of the marshy Sudd in southern Sudan. Eventually, sluggishly, and now known as the White Nile, it headed to Khartoum, where it joined the Blue Nile and then continued its northern progress through ancient kingdoms and into world history. Looking north and east, we were facing countryside ravaged by sleeping sickness and war; to the north and west lay an area that, in America at least, had become famous in 1999 as the origins of the West Nile virus.

The virus was first isolated in 1937 from a febrile adult woman in the West Nile district of Uganda; in the 1950s, Egyptian researchers found it in mosquitoes, birds, and people. In 1957, in Israel, it was recognized as a cause of severe human inflammation of the spinal cord and brain. Over the next few decades, cases in people and horses were recognized throughout the western Mediterranean, southern Russia, and eastern and southern Europe.

Like the geographically and emotionally complicated Nile River, which pre-dated its European "discoverers" by tens of thousands of years, the virus and the infection pre-dated its discovery by hundreds of millions of years. Some have now suggested that Alexander the Great may have died of West Nile virus; Babylon, where he died, was thick with mosquitoes, and ravens fell from the sky on the day of his death. Others believe that he more likely died of typhoid fever, a supposition that doesn't explain the falling ravens and to me is somewhat less interesting.

As long as West Nile virus kept its range to a variety of bird species, traveling regularly between Africa, Asia, the Middle East, and Europe, or to occasional human cases in marginalized Africans, researchers, at least in North America, didn't pay a lot of attention. To be fair, although there was evidence of the virus and some very closely related ones (such as Kunjin, which is found in Australia and some South Pacific islands) just about everywhere in the world, infection in either the birds or people didn't usually have serious consequences. But it is a sign of the general biological narrow-mindedness, the dazzling success, and the subsequent blindered vision of many scientists, that medical researchers didn't see the summer of 1999 coming.

In that hot, dry summer of 1999 in New York City, people (and a variety of birds at the zoo) were getting sick, and some were dying, from an unidentified neurological disease. At first, medical investigators called it St. Louis encephalitis virus. Actually, at first they didn't call it anything; the disease was hardly noticed at all. Who would notice a few people dying of strange diseases in New York City? When they did call it St. Louis encephalitis, it sort of made sense.

St. Louis encephalitis had periodically caused similar outbreaks in the southern United States and was, until West Nile virus came along, the most common cause of viral encephalitis in the U.S. In 1999, there were twenty reported cases of St. Louis encephalitis in New Orleans. That was about a typical year. In the mid-1970s, an epidemic afflicting about 2,500 people over thirty-five states was deemed by authorities to be an aberration. Another epidemic in Colorado in the mid-1980s infected over a thousand people, although fewer than two dozen reported being sick; one person died.

That pattern of sporadic outbreaks and silent infections, in which many people do not actually report being sick, is fairly typical for a lot of these viruses. Since the mid-1960s and until West Nile virus hit, there had been close to five thousand cases of St. Louis encephalitis in the U.S. Epidemiologists looking for patterns say that St. Louis encephalitis virus causes epidemics every ten or eleven years in the United States, apparently in inverse synchronicity with sunspot activity. I don't think public health epidemiologists were actively monitoring sunspots in 1999.

The 1999 suspect cases in New York were a bit out of the usual southwestern range for St. Louis encephalitis, but then so are a lot of species these days, including us, so this shouldn't be a surprise.

Mosquito-borne members of the extended family of the genus *Flavivirus,* to which West Nile virus belongs, may often only cause aches and pains, but they can, and do, lead to all kinds of neurological havoc in a small percentage of those who get sick. Kunjin virus in Australia is usually associated with a mild disease in people, but Murray Valley encephalitis, which cycles naturally between birds (herons and pelicans) and mosquitoes, is more often a life-threatening disease in people.

Japanese encephalitis, carried in pigs and wild birds in many parts of Asia, is spread to people through mosquitoes. Mostly this virus quietly cycles between pigs, birds, frogs, and mosquitoes, but in pigs it can cause abortion and piglet death. The type of mosquito that transmits the virus seems to reproduce especially well in fertilized rice paddies. In people, the virus may cause anything from a few aches and pains and fever to bleeding from the gums and nose to full-blown seizures and death. Tens of thousands of people get sick from it in eastern Asia every year. One in five of these die. There is a vaccine, which works pretty well.

Among all this family of diseases, the one most vigorously and seriously investigated has been yellow fever, probably because it was of interest to the U.S. military. In its home in tropical Africa, yellow fever cycles naturally between mosquitoes that breed in holes in trees and tropical forest primates. The disease itself can be quite awful. People vomit blood, and bleed from various other orifices; their skin turns yellow from liver damage. We might understand why port cities in the eighteenth and nineteenth centuries did not welcome ships attempting to enter port carrying sailors afflicted with what they called yellow jack.

Through our incorrigible meddling and bad luck, we not only managed to bring the virus to the Western Hemisphere but also managed to select for a species of the carrier mosquito, *Aedes aegypti,* that bred in holes in "horizontal trees," that is, in water tanks on slave ships. In tropical America, the disease caused considerable devastation for a couple of hundred years and settled into new niches, both urban and sylvatic. Yellow fever killed 27,000 of Napoleon's best troops in the Caribbean and opened the door for Haitian independence. A Cuban physician, Carlos Finlay, hypothesized that the disease was transmitted by mosquitoes, a finding that was confirmed by the U.S. army surgeon Walter Reed in 1901. The discovery of what spread the disease (mosquitoes) and how to control it (mosquito netting on beds, cleaning up standing water) gave the American troops an edge in their conquest of tropical America, including the ability to build the Panama Canal. Today the disease occurs as a human-mosquito-human disease in poor urban tropical areas, and as a zoonosis for those who have unprotected intercourse with certain tropical rain forests.

According to WHO, forty-seven countries in Africa and Central and South America, either entirely or in some regions, are endemic for yellow fever. In Africa, one study estimated that, in 2013, between 84,000 and 174,000 people came down with severe cases of yellow fever, and 29,000 to 60,000 died. Again, there's a good vaccine, and the fact that people are still getting sick and dying has more to do with social and economic justice than with a paucity of science.

Okay, so the deaths in New York were not yellow fever or Japanese encephalitis. And the West Nile thing seemed a bit of a zebra. But what if the hoofbeats you hear in New York City *are*, against considerable odds, those of a zebra? What if it is a virus from Africa and the Middle East? What if we live in the twenty-first century, when we are juggling and shipping and smuggling microbes around the world in agricultural produce and exotic pets and traveling relatives just about as fast as we can? Then you might be caught with your intellectual pants down.

After the genetic and antigenic mapping of viruses from a dead flamingo, a couple of people, and some mosquitoes, using envelope-glycoprotein-specific monoclonal antibodies (meaning they tested specifically for the proteins on the outside of the virus), the disease was confirmed to indeed be due to infection with West Nile virus. Not only that, but investigators could say with some certainty that it was related to one isolated from a dead goose in Israel in 1998 (ah, that I were so sure of my own relatives). By the end of that summer of 1999, more than sixty people were reported sick, and seven died, from the disease. A later investigation suggested that the number of people infected might have been over eight thousand. As is usual when we first stumble across a new disease (or it stumbles across us), we only noticed the most serious cases.

That initial outbreak of what turned into a continent-wide epidemic before settling into a source of chronic endemic disease and anxiety has spawned more versions and stories than actual cases of disease, which should tell us something about scientists, how science is really done, and how we might learn our way through the complex challenges of life in the twenty-first century. For West Nile virus in North America, as well as people suffering from serious neurological disease, there were ecological morality tales, whispers of plots and Jewish travelers, and great public health and biotechnological successes.

Some wondered what would have happened if paternalistic health officials at the CDC hadn't discounted the information from a (female) veterinarian that birds (Crows! Who cares?) were dying in New York City. By most accounts, Tracey McNamara, the veterinarian at the Bronx Zoo who was studying a substantial die-off of birds in the area, was apparently sidelined, ignored, and shabbily treated by the single-species-trained medical investigators even as she gathered the evidence needed to unravel the nature of the mystery agent. The number of cowboy egos riding around at the CDC—many since humbled by subsequent events—amazed some outsiders, but probably not those who are used to the Byzantine world of competitive science, inside and outside academia. In much of Canada, collegiality and teamwork tended to prevail over cutthroat competition, but, not to be too smug, much good scientific work in Canada has also been stifled by kowtowing to corporate or other ideologies and an intransigent refusal by many government departments to share information. Whether any of these shenanigans made any difference in the outcome is hard to say, since by the time the spraying for mosquitoes started, the epidemic was mostly over.

Some suggested that the West Nile virus outbreak was a bio-terrorist attack by a shadowy group of freedom fighters trying to bring down the Empire. Hadn't the most recently reported outbreak, killing about fifty of five hundred people affected, been in Romania? Wasn't there also an outbreak in southern Russia, before the crows started dying in New York? There was. As the political activist Abbie Hoffman once said, "Just because you're paranoid doesn't mean they are not out to get you." Or maybe the outbreak was just a message from our old friends the migrating birds, who return each year to remind us that we are part of a global ecosystem (which, as often as not, we then quickly forget). Certainly, the New York outbreak was in an area near wetlands frequented by migrating birds, mosquitoes, and people.

But once the virus was here in North America, it was here permanently, endemically. After that first summer, it seemed to lie low, getting a foothold: the right mosquitoes, the right birds, the right neighborhoods. Climate change had no doubt made North America more hospitable to a virus that must surely have visited our shores before: warmer winters allowing better mosquito survival, drought and heat bringing birds and mosquitoes closer together at pools of water (where they build up high levels of virus), followed by heavy rains that cause them to fan out over wide geographic areas. Whatever the reasons, the virus found a home, settled in for a year or so, then reproduced and headed west. In 2000, there were 29 human cases in the U.S.; in 2001, 666; in 2002, over 4,000; and the following year, almost 10,000. The numbers reflected a great wave moving across the continental United States and Canada. Behind that wave, the numbers settled, but periodically rose in waves again. In the U.S., there were 2,500 cases in 2004, and just under 3,000 in 2005.

In Canada, sick birds first showed up in 2001, and human cases started appearing in 2002, at which time Ontario's medical officer of health publicly dismissed West Nile as being trivial compared with the annual influenza epidemic that swept across

North America. This statement did not endear him to the worried public but raised some interesting questions. Should the importance of diseases be judged by the number of cases? Number of deaths? Disability-adjusted life years? The population affected? *E. coli* O157:H7 had affected many thousands of people with minimal social impact before it hit a few middle-class children, at which time it transformed the entire American meat inspection system. Are diseases that surprise us more important than diseases we have come to take for granted? Surprise, after all, indicates that either scientists are in the dark or they haven't been watching carefully enough in the right places. Surprise (more cases than "expected") is one of the key elements in an epidemic.

The first year, diagnoses were made slowly in Ontario because no one seemed prepared, although, based on tracking of possible paths the epidemic would take, investigators had a couple of years' notice. But the tracking came from those animal-disease people, and what do they know? This wasn't the flu, after all. Early reports suggested there were fewer than a hundred human cases that first year; later reports suggested closer to four hundred. In 2003, West Nile virus took flight westward across the country—or, more probably, having flown south from the eastern seaboard, the virus moved north on a new flight path. More than a thousand cases occurred in the Prairie provinces, and the disease spilled over the mountains to the West Coast. Birds, after all, fly, and many of the infected birds were returning from the southern United States, where the disease was already quite nicely settling into its New World habitat.

And everywhere, crows, ravens, magpies, and jays were dropping dead. Most of the dead birds were corvids, but not all of them. In Ontario, an outbreak in an owl rehabilitation center killed off most of the larger, northern-breeding owls but left many of the smaller, southern-breeding owls alone. In that colony, which was designed to rehabilitate injured birds and release them to the wild, the virus appeared to be transmitted by the biting louse fly. It is doubtful that the virus was transmitted to animals or people outside the colony. Most people tend not to get close to owls, but veterinary students can get weird that way.

When I was a student in veterinary college in Saskatoon, I spent part of a summer helping to retrain injured and repaired owls before they were returned to the wild. As part of the retraining, they would sit on my arm and then chase dead mice that someone threw. Partway through the summer, I started getting severe headaches, and my memory was affected. One day my sister-in-law telephoned, and, after a cheery but apparently strange conversation with her, I hung up and promptly forgot that she had called. When my wife figured out that something was wrong, she drove me to the hospital; I then spent many weeks in bed getting delirious, hallucinating, and suffering splitting headaches from what was diagnosed as viral meningitis. Had this not been well before the birth of Harry Potter, we would have suspected the owls of sending me messages from dementors, but it could have been a dozen other exposures.

Owls can carry *Chlamydophila psittaci,* the microbe that causes parrot fever. The organism used to be called *Chlamydia psittaci,* but the name was changed, in part to reflect new genetic information; the change also serves to clear up the confusion between this zoonotic chlamydial infection and the chlamydia that cause vaginal infections in people. Psittacosis in people can range from a mild respiratory infection to fever, chills, and serious pneumonia, but people more often pick that up from breathing aerosolized feces of pet birds. In the United States in 2005, there was an outbreak of salmonellosis among elementary school children who had dissected owls' fecal pellets. Despite these few possibilities, the cases of actual transmission are really few. During the peaks of the West Nile virus epidemic, dying or dead hawks, pheasants, and other birds that were suffering from the disease were also being taken in to wildlife disease investigation centers around North America.

Crows and ravens have historically been birds with messages, often dark ones. In this case, some people thought that the messenger might be the message, that crows were carrying the virus that was going to kill us all. A virus that kills so many of its reservoir host, however, wouldn't last long; this is Darwinian selection at its simplest. One continuing question in West Nile virus investigations is where the virus has been making its home when it wasn't out on killing sprees or instilling fear into suburbanites. Who *wasn't* it killing?

In 2004, Sharon Calvin, a veterinarian doing graduate studies under my supervision, decided to investigate a possible candidate for harboring West Nile virus: gulls. The Leslie Street Spit is the eastern headland of a three-mile-long peninsula extending from the Toronto shoreline into Lake Ontario. Now a home to a park and wildlife refuge, the peninsula has been built up from harbor dredge and industrial waste over fifty years. In early summer, as one stands at the edge of the spit before a heaving sea of squawking and yelling and crying birds amid flying feathers, with feces and urine spraying across the heat-edged sky, it is difficult to imagine a time that gulls were ever endangered.

As with so many other species, humans destroyed gulls' early abundance through destruction of habitat and use of feathers for hats and eggs for food. From huge numbers in the early 1800s, ring-billed gulls all but disappeared from the Great Lakes Basin by the beginning of the next century. Then, in 1917, Canada, the United States, Mexico, and Russia signed the Migratory Birds Convention Act. Those who doubt the effectiveness of such governmental agreements need only visit the Toronto waterfront. The agreement was aided by the provision of an abundant food supply for these garbage-eating omnivores and the creation of new nesting sites made of concrete, slag, mud, and other so-called wastes left in the wake of city building. Gulls, at least, benefited from this protection; there are now probably three-quarters of a million nests in the Great Lakes Basin and upper St. Lawrence River.

So, in 2004, there they were, multiplying as fast as we could feed them and provide habitat. And not dying from West Nile virus. We already knew that gulls, like Canada geese, another species to benefit from waterfowl protection, can pick up all kinds of gut-wrenching human pathogens from garbage dumps and farm manure (*Salmonella* and *Campylobacter,* for instance) and then spread them around to new places. We knew that bird (and bat) guano is an excellent medium for *Histoplasma capsulatum,* a fungus that, once inhaled, turns into a yeast and causes pneumonia in people, but that is a risk for people cleaning up old buildings, not for the general public. We knew that gulls can carry some strains of Newcastle disease virus, first discovered in Indonesia and England in the 1920s and now known to be a worldwide killer of domestic chickens. The virus is a member of the family *Paramyxoviridae,* which includes the viruses that cause measles and mumps, as well as the zoonotic Hendra and Nipah viruses (from bats); in its most pathogenic form, Newcastle disease has devastated poultry farms around the world. More recently, another strain has killed cormorants and pelicans in North America. In people (mostly veterinarians and poultry farmers), Newcastle disease causes conjunctivitis and headaches.

Being waterbirds, gulls also carry a variety of avian influenza viruses. An outbreak of avian influenza had killed a few hundred gulls living like vagabonds on a rooftop in my hometown, Kitchener-Waterloo, Ontario, and an earlier study had found evidence of avian influenza virus (H13N6, if you want the viral postal code) in almost all the 360 gulls tested from colonies on Lake Ontario. Well, waterbirds are the natural home of influenza viruses, so this should not have been a big surprise. Why not West Nile virus?

In the summer of 2004, residents of Toronto and Hamilton could see small teams of people in white space suits and hard hats (to protect them against dive-bombing, shit-spraying birds) chasing gulls with nets in the wild reaches of the Leslie Street Spit and the shores of Hamilton Harbour. Throughout the summer, Sharon worked with our team at the University of Guelph and the Public Health Agency of Canada, bleeding and swabbing mothers and chicks and trapping mosquitoes.

It turned out to be a great year for the public and a bad year to look for West Nile virus. We found evidence of the other two viruses Sharon was looking for, but neither the mosquitoes nor the birds came up positive for West Nile virus. It could have been the cooler summer, the government mosquito-control programs, the design of the study, or the tests we used, or it could be that gulls are not a big part of the West Nile virus story. That's science. Given that West Nile virus depends on mosquitoes and birds, it made sense, then, to ask about the influence of rural and urban landscape designs. Was there something about the way we had designed our urban areas that facilitated spread of the disease? The short answer was: yes.

West Nile and similar viruses, such as St. Louis, western, and eastern equine encephalitis viruses, are transmitted from certain species of infected passerine birds (such as robins and grackles, which serve as virus amplifiers) to some species of mosquitoes, and then to mammals, such as horses and people. Not all bird species are amplifiers: corvids, such as crows and jays, for instance, suffer high mortality and, while serving as early warning that the virus is circulating, are not important in its amplification and spread. Neither horses nor people appear to develop a sufficiently high viremia—the level of virus in the blood—to reinfect mosquitoes. From a landscape-design point of view, then, habitat for the relevant mosquitoes and birds invited closer examination.

More than twenty mosquito species have been implicated in the spread of West Nile in North America, but the mosquitoes that are of most concern have been *Culex pipiens* in the east and *Culex tarsalis* in the west. Both are active at dusk and dawn, spend most of their time in the tree canopy, and feed mostly on birds. Toward the end of the summer season, however, both species shift host preference and start to feed on mammals as well as birds. Both are well adapted to urban environments. *C. pipiens* prefers water that is high in organics, and is possibly the most pollution tolerant of all mosquitoes, being frequently found in sewage-treatment-plant effluent. *C. tarsalis* seeks out water with decomposing vegetation (roadside ditches, for instance), and has been known to develop its first batch of eggs without a blood meal in as little as four days after emergence. Both mosquito species are opportunistic container breeders and will lay eggs in nearly any standing water. Both species overwinter as adults and seek out humid, sheltered places for this purpose.

How do these ecological variables relate to landscape design? Storm sewers, which carry water runoff from residential and commercial properties, are often situated under concrete roadways. The sewers are warmed from above by the streets; this warming represents a localized example of the more general "heat island" phenomenon, in which concrete absorbs and reradiates heat. The heat island effect means that the downtown areas of cities are generally warmer than surrounding green spaces. Being semi-stagnant and underground, storm sewers are also free of insect predators such as fish. All these characteristics make storm sewers an ideal habitat for breeding, and overwintering, of certain species of mosquito. Furthermore, many urban and suburban landscapes that provide important living space and fresh air for people also offer excellent habitat for the bird species, such as sparrows, finches, and grackles, which are the most competent hosts for the virus.

A few studies by ecologists have concluded that populations of plants, mammals, birds, and insects living in ecosystems with low biodiversity tend to be more adversely affected by host-specific disease, and more effective at spreading it, than populations in ecosystems with high biodiversity. Because potential host populations in these low-diversity areas are less constrained by competing species and predators, the hosts tend to be more densely distributed, and more likely to spread disease among themselves. This "dilution effect" of high-biodiversity ecosystems appears to hold true for Lyme disease transmission in the United States; is it also true for West Nile virus?

In Sweden, researchers found a direct relationship between the amount of biodiverse green space (forest and preserved natural ecosystems) and bird populations in urban areas. Species diversity for woodpeckers, hole nesters, and forest birds increased from the city center to the periphery, along with vegetation diversity. However, urban birds were concentrated in the residential area. While the residential habitat supported large numbers of birds, the species diversity was quite low, suggesting that while the habitat was good for specific urban birds, it was hostile or unsuitable for many other species. The researchers posited that the green wedge of the greenway increased species richness for nearby areas, and they concluded from the data that increasing the number of greenways into the city should increase the species diversity for the entire area, including the city center. Vegetative groundcover was also positively correlated with species richness.

The Swedish study of avian biodiversity in cities is especially pertinent to West Nile virus. All of the ten most virally competent birds in the United States and Canada could be qualified as "urban birds," based on their prevalence in annual backyard bird counts. Jays, robins, and grackles are common backyard birds across North America, while crows and magpies are common urban scavengers. Both finches and sparrows are also ubiquitous in urban areas. While many of these birds migrate in the wild, their urban counterparts often stay all winter, because their food sources (backyard bird feeders, roadkill, and dumpsters, to name a few) remain.

What ties many of these birds together is their need for clear ground to forage on and large trees for nesting or roosting. The mature tree canopies and open lawns found in many residential environments, when taken together with stagnant storm sewers and concrete-generated heat islands, create a high-risk environment for the maintenance and spread of West Nile virus as well as other, similar viruses.

After a great wave of money, panic, and scientific work, public interest in West Nile virus subsided. The virus is here, quietly circulating in its newfound home, periodically making non-human animals and people sick, sometimes tragically so.

Fifty years ago, many North Americans dreamed of green lawns and wide streets. Wasn't "nature," however trimmed and coiffed, a good thing? If, in West Nile virus, nature was reminding us that not all green spaces, birds, and mosquitoes are alike, that there might be a downside to suburban green lawns and broad streets, have we listened? Are we paying attention? Do we welcome the cacophony of crows in spring?

If we brushed aside the quiet rustling voices of the habitats of our homes and gardens, we need not worry. Nature would speak again, and often, until we heeded. In 2018, for instance, the CDC reported 2,647 cases of West Nile virus and 167 deaths. In Canada, after rising in 2003 to 1,481, and peaking at 2,215 in 2007, human cases dropped to dozens or at most hundreds each year; in 2018, 427 Canadians were reported to have the disease.

As the experts at WHO have already noted, there is no non-pandemic period. Tread lightly.

(9)

THE CHICKENS FIGHT BACK, WITH DUCKS

When, in 1997, Lam Hoi-Ka, a previously healthy three-year-old boy, died of multiple organ failure in Hong Kong, no one suspected what a world-changing event this would be. When a team of virologists from the Netherlands declared that the death-dealing agent was H5N1, a virus that was previously known only to infect birds, scientists were shocked. Every year people around the world came down with "the flu," but not this one, and never a virus directly from chickens. The theoretical possibility of a deadly global pandemic, similar to the 1918 flu that killed millions of people, was suddenly made real.

It is worth trying to remember how different the world was in 1997. Few people anywhere in the world owned personal computers. Even in 2003, when SARS first made headlines, Facebook was essentially an online college locker room where boys rated girls, Twitter had not yet been invented, and the idea that major news organizations might go online was treated by many in the news business with derision.

All this is to say that when bird flu first catapulted to world attention, most of us got our news from newspapers or television.

While virologists and epidemiologists from around the world urgently tracked down the origins and initial spread of the new virus, most of the rest of us watched the television images of people walking around in white body-covering suits designed for dealing with hazardous materials (hence: hazmat suits). They looked a little like the outfits beekeepers wear, except that these workers were carrying away and dumping garbage bags full of dead chickens.

Influenza viruses are simple, tiny things: eight segments of ribonucleic acid, or RNA, packed into a string or sphere about 80 to 100 nanometers in diameter (think billionths of meters; a hair shaft is about 100,000 nanometers, a bacterium about 1,000). Being small, and not, technically, alive, they can get away with stuff bigger things cannot. For instance, they can commandeer the machinery of a living cell to reproduce. Once inside, they are sloppily promiscuous, sharing bits and pieces, it seems, with whoever happens to be around. These viruses, classified as types A, B, C, and D, circulate in people, pigs, horses, birds, mink, cats, whales, and seals.

As with personality types in people, type A is the most worrisome to health workers. Type A is subdivided according to their HA (hemagglutinin, of which there are at least sixteen types) and NA (neuraminidase) antigens (projections that stick out of the lipid envelope surrounding the RNA; there are about nine types of these). In chickens, a particular subtype of influenza A causes disease known historically as fowl plague; there are low pathogenic avian influenza (LPAI) strains and highly pathogenic avian influenza (HPAI) strains. HPAI can kill a lot of chickens quickly.

The many different combinations of the influenza virus genes vary almost infinitely in their ability to cause disease, which depends not just on the virus itself but on the genetics, health, nutritional status, and organizational structure of the populations into which they are introduced. The combination that was creating the greatest anxiety around the world in 2006 was H5N1 (hemagglutinin type 5, neuraminidase type 1); although it is not easily transmitted, when it does transmit to other birds, to cats, to people, and to a variety of other species, it is unusually lethal. Influenza viruses were not supposed to behave this badly.

What made the 1997 news images confusing and alarming were the stories that accompanied them. In Hong Kong, a chicken-adapted virus had jumped directly to a person. This hadn't been reported before.

In Canada and elsewhere, people in hazmat suits were disposing of chickens that were said to be infected with LPAI, which was said to be not of concern to humans. But why, then, were they wearing those white suits? What wasn't made explicit, at least as I recall, was that LPAI viruses can mutate into HPAI variants rather quickly.

For scientists, politicians, and just plain folks, this was surreal. None of us had a satisfying story that would explain, and hopefully give some reasonable context to, the images. For most people, chickens were "healthy, low-fat" meat you bought in plastic packages at the grocery store, not agents of mass death.

As a veterinarian and epidemiologist, my own chicken stories were different from those of many North Americans and Europeans, but offered equally unsatisfying explanations.

Between 1985 and 1987, I was on a contract with the Canadian International Development Agency near Yogyakarta, in Java, Indonesia. While I was working to set up a diagnostic laboratory for animal diseases, my wife, a nurse with a masters in health sciences, home-schooled the kids and worked as volunteer in community health.

Chickens were everywhere.

The bright, fierce, piercing cry startled me from sleep. Groggily, bathed in sweat, I pushed away the long, cylindrical pillow so beloved in all traditional Javanese hotels, which my Indonesian friends assured me was called a "Dutch wife." In one stumbling motion, I swung out of bed, pulled my sarong around my waist, ran to the door, and stuck my head outside into the misty tropical air. Several yards away, on a low branch, perched the regal father of all chickens. His wattles were full, fleshy, and pink. Like the robes of some strutting bird-king, his cape flowed down from his crown, over his back, in brilliant bronze, rust, and gold, ending in a ruff of white. His lower chest and fountaining tail feathers shimmered in the early morning sunlight from deep black to blue to green. His sharp black eye caught mine and held it. This was not a bird to quietly back down from a challenge.

The red jungle fowl, and its gray and green siblings, was first domesticated in South and eastern Asia—India, Myanmar, Thailand, Laos, Vietnam, China, Malaysia, Java—many thousands of years ago. These birds were originally bred for cockfighting. From India, the birds explored westward into Persia, carrying with them their wild spirit, and, one assumes, the populations of viruses and bacteria living in their intestines. There is no evidence that they were considered a low-fat, low-cost, low-carb food or a public health threat.

The Zoroastrians praised cocks for driving out the devils of night and guarding the household. From Persia, the birds invaded both northern Africa and Europe. Although the "rosy-fingered Dawn" Homer described in *The Odyssey* was not accompanied by cockcrows, Aeschylus (late in the fifth century BC) has Athena (in the play *The Eumenides*) warning the Greeks that civil war is like cockfighting, an image (fierce, small-brained, beautiful birds passionately slashing each other to death) that some of us would extend to war in general, although cockfighting itself has perhaps more to recommend it.

The Romans used live chickens for augurs; if the birds eagerly ate food thrown to them, things would go well. If not, this was a poor omen. As might be expected, the chickens were kept underfed until omen-reading time came around. During a 249 BC battle against the Carthaginians at Drepana, in what is now western Sicily, the Roman consul P. Claudius Pulcher threw the sacred chickens overboard when they refused to eat, saying, “If they will not eat, let them drink.” Later the same day, the Carthaginians sank 93 of the consul’s 123 ships. The consul was promptly recalled and forced to pay a large fine. One could draw a lesson from this event, but it would be premature for me to pronounce omens so early in this story.

Later on that 1986 morning in Java, sipping my jasmine tea and nibbling on sweet buns, I watched a scattering of various-sized *kampung* (“village”) chickens scratching through the dirt between the buildings, under the banana, mango, and jackfruit trees, skittering out from under the wheels of predatory buses and trucks roaring by in clouds of dirty diesel on the nearby asphalt highway. A woman leaning forward on her motorcycle, a feathery bustle of chickens hung by their claws and flouncing out behind her, snarled past, heading to market. The rust and black feathers of the *kampung* birds, although thin and shabby by comparison with those of their wild cousins, still showed traces of royal blood.

In the late 1980s, in the local markets of Java, *kampung* chicken brought a premium price compared with the amount paid for their fast-growing, placid urban cousins. So precious were they that nothing was wasted. Our family dined at Nyonya Suharti’s, Java’s answer to KFC, in Yogyakarta, in the province of Central Java. The pieces were tasty, very tender, and finger-licking greasy.

“You know,” our nine-year-old son said, “in this country, some people eat the chicken heads and everything.”

I already knew about, and had tried, the curried soup made with chicken intestines. I looked at the batter-smothered piece of bird I had just set down and recognized a beak, a comb, and an eye. The brain was already eaten away—by me. Everyone at the table laughed uproariously at my lack of attention to what I had been eating.

"That's something to write home about," laughed my wife, Kathy.

"The brain tasted like liver," I said.

Kathy looked down at her plate, at the piece she had just finished: a beak, a comb, an eye, the brain eaten away.

"I wondered what that was," she said.

Eating the brains of chickens does not have the same moral or infectious-disease baggage that eating monkey brains does. Still, even for someone who was, when faced with some straggling bits of dinner left on the plate, reminded about the starving children of India, the idea that some cultures really do take to heart the adage "Waste not, want not" can be disconcerting. And if there was no waste at the dinner end, neither was there waste at the farm end. Unlike their urban cousins, the village birds were fed nothing except scraps, required no extra building investments, and provided for their human neighbors' food, entertainment, and a sense of wonder at the beauty and cleverness of these silly birds.

Sometimes, as I drove through the country for my veterinary epidemiological work, I would pull over beside the narrow asphalt road to watch a line of ducks following a thin man, who carried a white flag flapping atop a bamboo staff. The man would have started from East Java, so I was told, with tiny ducklings. They would be imprinted on the white flag, following it like the dutiful, small-brained beauties that they are. The duck herder would stop at the watery muck of a rice field, where the ducks would forage all day, acting as pest controllers. At night, they would gather in a close flock around the white flag, for sleep and protection. The next day, they would walk on to another field, farther west.

Walking or swimming in roadside ditches, foraging in rice fields, they would make their way to Jakarta, far to the west, where the ducks would be sold. The flock leader would take their money, ride the train back to their home in the east, and start over. Not a bad life, I thought, for the ducks and their adopted parent. Good ecologically, good economically. And good, it turns out, for at least some of the viruses and bacteria that make their homes in bird poop.

A few years later, I stood in a broiler barn, that is, a barn where chickens are grown for meat, in southern Ontario. In my plastic boots and white throwaway safety suit (to protect the birds, not me) and a face mask (to keep the fine particles of litter-dust out of my lungs), I gazed out over ten thousand identical birds. The room was spacious and the litter was clean. Delivery of food, water, and air was computer controlled. The birds were white-feathered, plump, and only mildly curious about life. These were urban, office-dwelling birds. In five weeks, they could grow—uniformly, all of them—to the exact size required by KFC.

By the wonders of genetics and intensive breeding for specific traits, the fiercely wild stock of the jungle fowl had been transformed into something that could grow faster, more uniformly, and, by some standards, more efficiently. And if the genetic stock, the feed, and sometimes even the buildings were imported, and underpaid local labor used in place of underpriced fossil fuel, this feat could be accomplished just about anywhere in the world. This, too, is something amazing.

Between 1961 and 2017, world poultry meat production increased from 9 to 122 million tons, and egg production shot up from 15 to 87 million tons. Since most of us experience a sort of cognitive dissonance when we see chickens and tons in the same sentence, let me rephrase this. In 1961 there were just over 3 billion people and just under 4 billion chickens in the world. In 2020, as I write this, about 7.7 billion people are jostling and shouting for space here, along with more than 20 billion chickens—and perhaps as many as 50 billion if one considers the short slaughter-and-restock turnover of those populations.

The fastest growth in commercial poultry production has been in the developing world. By the late 1990s, countries such as Indonesia and Brazil were increasing their commercial production by about 10 percent a year. When I was visiting South Sudan in 2012, just a few months after it gained independence after thirty years of civil war, I found "fresh" Brazilian chicken for sale in the market in Juba, the capital. China, already one of the world's biggest producers, was increasing chicken production at about 4 percent annually in the 1990s. Chickens were being grown, trucked, shipped, and fried as fast as the technology allowed. Who would have thought that so many people on this planet could be fed with such apparent ease?

Well, the geneticists helped. These birds feeding the world weren't just any old chickens. In 2018, the authors of a peer-reviewed research paper declared that the "skeletal morphology, pathology, bone geochemistry and genetics" of modern commercial chickens—whose global body mass now exceeds that of all other birds combined—are so different from their ancestors that they may be considered a "novel morphotype" symbolizing "the unprecedented human reconfiguration of the Earth's biosphere."

Who would have thought human beings were so clever, and so naive? But in ecology—which is to say, in a world where everything is, sooner or later, connected to everything else—there are costs, and trade-offs. Who would have thought that this feat would also create the perfect conditions for barn-sized outbreaks—thousands of uniformly susceptible animals gathered under one roof—and then fit so neatly into a globally integrated system that created perfect conditions for a pandemic? By the end of the twentieth century, epidemiologists who specialize in food-borne diseases were already well aware that a pandemic of salmonellosis, a disease with both immediate effects on the gastrointestinal system and long-term effects on arthritis and cardiovascular disease, was one of the hidden costs of mass producing chicken (see my book *Food, Sex and Salmonella* for more on that).

This pandemic could have been taken as a warning, an omen from the chickens of the world, a shot across the bow, as it were. The omen was not cryptic. It might have been something like: chickens carry their own bacterial and viral microbiomes; the economies of scale for chicken production are the same as the economies of scale for disease; small farms have outbreaks; big farms breed epidemics; globalization of big farms creates pandemics. But, like Pulcher 2,200 years previously, humanity ignored the omen, threw the birds overboard, and put its faith in advances in food safety technology.

The omen having been ignored, few researchers in the late twentieth century would have stood in a chicken broiler barn and made the mental connections to ducks flying overhead, or to the populations of viruses and bacteria and parasites to whom the ducks are a quiet, comfortable vehicle for long-distance air travel. A few North American veterinarians, on thinking about ducks, might worry about them carrying the Newcastle disease virus that affects poultry, or the schistosome parasite that causes swimmer's itch or duck itch in people. These schistosomes normally cycle from waterfowl through open water to snails and back out through open water to birds or muskrats. Unlike their cousins *Schistosoma japonicum* in Southeast Asia, which can cause serious harm, these parasites don't actually like people. Their dives into human skin cause an allergic reaction and are kamikaze nosedives into itchy oblivion. But I digress.

Not long after visiting that secure, clean, broiler barn in Ontario, I stood outside in crisp fall sunshine next to the riffling waters of a beautiful Ontario marsh with a group of veterinary students learning about ecosystem health. In the crook of my arm, nestled on her back, was a wild mallard duck. I was swabbing her cloaca, the common passage where urine and feces come out. We were checking her for influenza virus, as part of a mock influenza outbreak.

Waterbirds such as geese and ducks were domesticated even before jungle fowl, probably in the "cradle of civilization," the Middle East. Their intestines are the natural home for all sixteen known subtypes of influenza A viruses. You can test waterfowl of all sorts—ducks, geese, gulls—just about anywhere in the world and find some variations of them. Sharon Calvin found evidence of influenza infection in 80 percent of gulls in Toronto and Hamilton; many of the ducks we swabbed at that marsh as part of the mock influenza pandemic also harbored influenza viruses. It would have been surprising not to find them. For the most part, these influenza viruses in waterbirds are evolutionarily adapted and don't cause any disease problems in their natural hosts. However, when waterfowl are mixed with chickens and pigs and people in close quarters—as my ecologically friendly farmer friends were doing in Southeast Asia—novel opportunities for the viruses are created, they become genetically more unstable, and evolution is accelerated.

In 1996, a precursor of the H5N1 virus killed some geese in southern China. No one paid much attention. Then the virus picked up some gene fragments from quail and ducks, spread to the poultry markets in Hong Kong, and made the leap to humans; it killed six of eighteen people who were infected. Mass killing of all the domestic poultry in Hong Kong temporarily stopped the problem, but the virus continued to infect ducks and geese and to happily, sloppily evolve. In late 2002, a new variation of the virus killed off most of the waterfowl in Hong Kong nature parks. In the next few years, the new, more lethal variant spread through Vietnam, Thailand, Indonesia, Cambodia, Laos,

China, Malaysia—the whole regional market. Not only was it making birds sick and killing them, but it was also infecting cats and ferrets and, finally, people.

When I lived in Indonesia in the 1980s, farming appeared to be an agro-ecologist's dream—lots of diversity, recycling of resources, low energy inputs. Influenza was a disease of people that periodically emerged from the creative microbial populations of southern China and wafted around the world, killing (mostly) older, already-debilitated people but generally leaving both non-human animals and the physically fit of our own species unscathed. Biodiversity in agriculture was seen as a solution, not a problem. So what happened?

Many commentators on the twentieth-century spread of influenza refer to the global pandemic of 1918, which started with American troops in Kansas and then spread around the world, but the stories go back much further than that. Hippocrates recorded an epidemic in 412 BC, and the name "influenza" itself comes from fifteenth-century Italy, referring to unexpected epidemics believed to be under the influence (influenza) of the stars. Not surprisingly, given the origins of chickens themselves, the first epidemic for which there is reasonably good information appears to have originated in Asia, before spreading to Africa and Europe. Ducks and chickens have been hanging around each other for a long time, and some kinds of influenza viruses have caused problems for as long as we can, collectively, remember. We learned to adjust; we monitored where the viruses came from and created vaccines. For most of the twentieth century, these practices kept matters more or less under control.

In May and June of 2005, one of the new variants of H5N1 killed more than five thousand wild bar-headed geese, gulls, and ducks in Qinghai Lake, China. Before they died, the affected birds had trouble standing and developed neurological problems (flopping around). Many researchers were worried that migratory birds would carry the virus down flyways into India. It looks as if they may have indeed carried the virus to Europe and Africa, but the evidence will always be ambiguous, and the world trade in poultry, both official and unofficial, remains an alternative explanation for how the disease has spread. To transport the virus, the wild birds have to be alive; if it is killing them, they stay where they are. If they survive, it could be because they have developed immunity or because the virus has evolved into something less pathogenic. In August 2006, the same virus turned up in cats in northern Iraq; so, however it was being transported and however dangerous it might be, it was certainly off to see the world.

Migratory birds carry all kinds of influenza viruses around the world, but the strains they carry usually don't kill them. If they are dying, it is likely because they are picking up new strains, which have evolved in domestically farmed populations. Another likely means of dissemination for the viruses we fear most is through the trucking of domestic and exotic birds around various continents—in other words, by people.

Some methods of spread are very culture-specific. In Thailand, fighting cocks are highly valued, reminding us of the origins of these birds as fighters. At matches, those betting money may go around and blow into the beaks of the contenders to check for lung capacity. Owners may suck mucus out through the beak between rounds. Traveling from match to match around the countryside, combined with this intimate contact, makes the fighting circuit a prime means of spreading the disease.

In October 2004, a Thai man flew into the Brussels airport from Bangkok with two crested hawk-eagles (*Spizaetus nipalensis*) in his hand luggage. He told customs officers they were for a friend. The birds, wrapped in cloth, stuffed into wicker tubes, and kept in the bag with the zipper left partly open for air, seemed healthy. They were "humanely sacrificed" (a very appropriate use of the term, since they gave their lives for the perceived greater good), and scientists looked for Newcastle disease and avian influenza. They found not only influenza but also the H5N1 strain that everyone feared. This kind of illegal trade is a huge business and a marvelous opportunity for all kinds of bacteria and viruses to find new homes in new lands.

Millions of tons of biological material—human food, animal feeds, meat-and-bone meal (MBM, which is basically animal offal reinvented as fertilizer in Indonesia, supplements for cats in America, and a variety of other essentials of modern life)—circle the globe annually. Economically, this circulation is described as free trade. Scientifically, it is simply carrying various types of living and non-living things from their ecological homes, where they usually cause few problems, to new areas, where they are very likely to cause a lot of problems. MBM was a major vehicle for transmitting bovine spongiform encephalopathy (BSE, or mad cow disease) within and between countries. One might be curious to know whether those who carefully crafted the documents of the World Trade Organization considered this possibility. One might wonder how many of them had some rudimentary knowledge of biology or even knew where food came from.

The farmers of Southeast Asia didn't scale up their production and increase the volume and speed of their trade in poultry products just "because"; they were responding to market demands for low-cost animal food. They are making a living by providing the protein and other nutrients that are necessary to sustain urban life; those farmers are, in a real sense, an essential part of the urban ecosystem. Without rapid economic growth and urbanization, avian influenza would likely remain a minor problem.

According to the United Nations, one in three people lived in a city in 1960. By the end of the twentieth century, almost half of all people did; by 2030, more than 60 percent of the population is expected to live in cities. Many of these megacities are in the developing world, especially in South and eastern Asia. These urban people want to eat, and they want protein. Chicken, ducks, geese, and pigs, all grown on an industrial model, will do just fine. Brazil, the United States, China, and the European Union are the world's biggest poultry producers. China leads the world in ducks and geese. Not surprisingly, other countries in the region—Thailand in particular—wanted to take advantage of these expanding urban markets and jumped into the hot economic fray. In places such as Thailand and Indonesia, increased production has sometimes been achieved by taking the kinds of laid-back, no-input chicken rearing I saw in the 1980s and scaling it up. If some chickens and ducks, mixed with rice paddies and fish ponds, are good, why aren't more better? Not surprisingly, the current epidemic of avian influenza started in south China and has spread outward from there.

Still, the temptation to blame the wild birds, and not our own hubris, is strong.

When H5N1 appeared in Russia, rumors circulated that hunters were offering to shoot ducks to prevent the spread of the disease. Westerners laughed at this primitive response.

In October 2005, when cases of H5N1 influenza started appearing in Europe, British hunters stood at the ready, assured that shooting ducks was no mere macho hobby but an act of national public health importance, a serious form of scientific surveillance.

That same month, George W. Bush, the president of the United States, gave the only response he seemed capable of imagining, saying he would rally the armed forces to fight the dreaded bird flu.

People who study viruses have pointed out that the H5 type of influenza virus found in China and a highly pathogenic H5N1 isolated from poultry in Scotland in 1959 share a common ancestor. Because influenza viruses are constantly drifting (small changes) and shifting (large, abrupt changes), the currently circulating strains are different from those first isolated. However, it would be disingenuous to suggest that intensification and global genetic "homogenization" of poultry production have not been driving forces in a variety of epidemics, including this one.

Some pontificators have suggested that farmers in South and eastern Asia should raise chickens the way we do in North America and Europe—inside tightly controlled buildings. These people have never lived in poor countries in the humid tropics, nor do they understand the systemic ramifications of creating a few large farms where once there were many small ones. If they want bio-security such as we have in Europe and North America, the tropical farmers will need to close off the barns. But in the tropics, without air conditioning, the birds will start to die within minutes. With what power source would they air-condition? And what would happen to all those poor farmers in the countryside who depend on small flocks of poultry for food and to pay their school and medical bills? In the 1980s, when we marveled at the man with the white flag and his ducks, the village chickens, and the wonder of the wild jungle fowl, were we off base? I think not. We've learned some things since then, and one of them, surely, must have something to do with the bewildering complexity of the world we live in.

In 2005, the Millennium Ecosystem Assessment, an unprecedented global scientific effort to assess the world's ecosystems, demonstrated all the services that intact ecosystems provide for us—air, water, food, meaningful work—and just how vulnerable we all now are, with water and soils and forests and oceans at or near tipping points. No, if we want to find a solution to the influenza problem, more of the same won't cut it. Even if we kill all the sick chickens and put the rest into air-conditioned hotels, there will still be ducks flying overhead or cats or ferrets slinking in and out of the shrubbery. Some of the largest outbreaks of avian influenza have been in some of the best-managed poultry operations in the world, in some of the wealthiest countries. The viruses, like all microbes, adapt quickly to new situations. Already in late 2006, a new variation of the H5N1 avian influenza virus had emerged in China, possibly in response to selection pressures from vaccines given to chickens. By 2016, human cases of H5N6, H7N9, and H9N2 were reported.

People in cities need food, but they need to be educated about where food comes from and the real costs of producing, processing, and shipping it. The energy costs of commercial chicken production, for instance, cannot be simply reduced to intake over output and then compared to cows, as is sometimes done. In this context, chicken always appears more efficient than other animal meats. But how much energy is required for the buildings and machinery, for growing special feeds, for getting the feed to the farm and the chickens to market? If the small farmers lose chickens as a source of livelihood, how will they pay for education and health care? All of these are real costs, which someone is paying.

Knowledge of the social and ecological dimensions of food should be part of every food consumer's education. An inability to shop for food, prepare meals, and talk intelligently about where that food comes from should be grounds for dismissal of politicians and corporate heads. In the years following the initial outbreaks of avian influenza, I spent a lot of time and energy working with policymakers in Canada, looking at how to prevent the disease from entering North America, and farmers in Southeast Asia, looking for ways to stop the epidemic at the source. Many officials and corporate leaders were encouraging countries to follow a program of test-and-slaughter, and of discouraging villagers from raising free-run village chickens. In March of 2008, at a market in eastern Thailand, I discovered that if sellers were responding to economic incentives, the programs designed to stop people from raising backyard chickens were unlikely to succeed. According to a woman with a dozen gutted and cleaned birds in front of her, village chicken was going for about twice as much on a per-weight basis as the commercially-reared broilers.

Many of those who were successfully lured into subcontracts with large commercial firms were men. This repeated a pattern often seen in the history of economic development: when an activity is for home subsistence (backyard chickens, egg money, home-reared crickets) the women are in charge. As soon as there are economic returns in the market place (industrial farming, competitive roosters) the men take over. Every solution to feeding the world and preventing pandemics has stumbled into this bramble of patriarchy and, in "solving" the original problem, creates many others. Every food security and pandemic prevention initiative, if it does not address the issue of gender relations, is, or will be, a failure in enabling sustainable human well-being.

The next month, at the invitation of my Indonesian co-workers, my wife and I visited a Javanese village in an area reported to be highly endemic for avian influenza. Nevertheless, villagers who attended our workshop claimed to have had no confirmed cases of bird flu. Birds had died of other causes, of course, and the carcasses were "safely" disposed of in a nearby river.

After the meeting, the villagers took us to see their chickens. Their greatest source of pride were their Ayam Pelung—competitive singing roosters. They were tall—about three feet high—and their calls were long, drawn out, low voiced, reminding me of Cesária Évora, the "barefoot diva" of Cape Verde. These singing roosters were each worth US$2,000 to $3,000, which was more than the annual income of most of these farmers. Before avian influenza hit the headlines, the roosters were taken to competitions throughout the country, and beyond. Winners could take in $500 in a single show. One of the farmers was a breeder who had sold roosters to buyers from as far away as Japan. A program that relied on slaughtering chickens that tested positive, and paying compensation at market rates for commercial broilers, was a non-starter for these villagers. Of course, they had no bird flu. Of course, any birds that died had succumbed to some other disease. Those who were designing global programs to eradicate avian influenza may have imagined that a chicken was "just" a chicken, but this was clearly not the case.

I remembered, then, seeing the roosters in competition back in 1986, in cages, high up on swaying poles, judges moving from bird to bird, listening. I do not know what criteria the judges used, but standing there in the slight cool of a shady tropical morning, hearing those *fado*-like songs of love and loss, I could understand, like the Zoroastrians, how the calls of those jungle fowl might drive away the devils of darkness.

(10)

QUEENS OF THE SOUTH: NIPAH, SARS-COV, AND SARS-COV-2

NIPAH

In the first week of October 1994, I received an electronic message from ProMED, at that time a fledgling international electronic bulletin board on emerging infectious diseases. Field workers from all over the world were encouraged to post new or unusual disease occurrences, and others were invited to provide insight or advice. ProMED has since grown into a large, effective international network with instantaneous reports of new or emerging diseases from even the remotest parts of the planet. It is the scientific community at its best.

The posting that day was a letter from the Australian chief veterinary officer to the director general of the Office International des Epizooties in Paris. In it, the veterinarian described the deaths of fourteen horses from an acute respiratory syndrome between September 7 and September 26, 1994. The horse trainer also died of a similar acute respiratory infection, on September 27. The agent was unknown.

The newly identified virus eventually took on the name of the suburb of Brisbane where the outbreak occurred—Hendra. A dozen horses, one man: one could say it was not something to lie awake at night worrying about.

In late 1998, also on ProMED, reports came out of Malaysia about an outbreak of Japanese encephalitis, caused by a virus related to West Nile virus and transmitted by mosquitoes. Since Japanese encephalitis is endemic in that part of the world, these reports were interesting but not terribly surprising. By December, eleven people had fallen ill and four had died; farms were fogged with insecticides, people were vaccinated, and the outbreak was declared over. By February 1999 there were twenty-five cases, with thirteen deaths; in mid-March, it was reported that ninety-eight people had fallen ill, forty-four of them fatally, within a three-day period. Thousands of homes and pig farms had been fogged with insecticides and thousands of people vaccinated with a Japanese encephalitis vaccine. Later the same month, there were reports that abattoir workers in Singapore had come down with encephalitis after slaughtering pigs from Malaysia. Japanese encephalitis is transmitted by mosquitoes. How could this be Japanese encephalitis? Was this, in fact, not Japanese encephalitis but a new disease?

Shortly afterward, the disease was being described as "Hendra encephalitis," and later, because it first appeared in the village of Sungai-Nipah, as Nipah virus encephalitis.

A new disease, characterized by fever and coughing followed by drowsiness and sometimes a coma, seemed to have come out of nowhere. It seemed to start in pigs and spread from there to dogs and people; older animals were more likely to become severely ill, with unsteady gait and frothing at the mouth. Scientists and investigators from around the world joined forces to investigate.

Reports from the CDC investigators sounded both excited and anxious. This virus had never been seen before. No one knew how it was transmitted or what it was capable of doing. Scientists wore gowns, gloves, and battery-operated respirators as they slogged their way through the pig farms at the center of the epidemic.

Even as the scientific investigation proceeded, the Malaysian army came in. More than a million pigs were slaughtered and buried. By the time the epidemic was over, in May 1999, 265 people from Perak, Negeri Sembilan, and Selangor states in Malaysia—mostly workers on pig farms—had fallen ill with encephalitis; more than a hundred died.

Where had the disease come from? Because pteropid bats (flying foxes) were suspected of being the main reservoirs of the related Australian Hendra virus, investigators pursued that avenue as one of several leads. Flying foxes (*Pteropus* spp.), which have a brain structure similar to that of primates and flying lemurs, are an interesting bunch. They are megabats, classified as megachiroptera to differentiate them (as if we couldn't tell) from microchiroptera, the microbats. Megabats have keen eyesight and roost in trees rather than in caves or houses. Some of the males sing in courtship. The females have single babies and nurse them at breasts. They eat fruit and nectar and are important for the survival and spread of flowering and fruiting plants in Africa, Asia, and Oceania.

Bats of various species are home to a variety of interesting viruses, including Hendra, Nipah, and, as we subsequently discovered, SARS-COV and SARS-COV-2, as well as those that cause Ebola and Marburg hemorrhagic fevers in Africa. Not only that, but a group of researchers from California State University have proposed that the Chamorro people of Guam, who have one of the world's highest rates of a fatal, progressive paralysis, amyotrophic lateral sclerosis, better known as ALS, or Lou Gehrig's disease, are being poisoned by a neurotoxin. This toxin, which comes from the fruit of the palm-like cycad *Cycas micronesica,* is concentrated in, among other places, flying foxes, which the Chamorro cook up in coconut milk.

Megabats are important for the ecological sustainability of the planet. They are gregarious, and they are excellent social role models for children. They apparently taste pretty good, too, and are sold as food in many east Asian markets. And they are potentially deadly and not to be messed with. A lot like us, then.

But how would the virus get from fruit-eating bats to pigs and people working on pig farms? And why in 1998–99?

A plausible story that has emerged from the investigations brings together a complex set of interactions. Sadly, the story is both messy and increasingly common.

The El Niño–Southern Oscillation (ENSO) phenomenon is a huge natural cycle of oceanic and air currents across the southern Pacific. It was called El Niño by the Peruvians because it arrives just after Christmas, *El Niño* being the Spanish term for the Christ child. ENSO has been related to a wide variety of natural events, from the upwelling (and disappearance) of anchovies in Peru to monsoon rains and droughts in South and eastern Asia and all the way over to the east coast of Africa.

Scientists think that the ENSO cycles have been getting more frequent and more severe, probably as the result of human-induced climate change. The 1997–98 ENSO was the worst on record to that date, resulting in a major drought through Southeast Asia. Even as the drought deepened, Indonesian forestry companies slashed through the forests of the area with wild abandon, and desperate Indonesian and Malaysian farmers set fires to clear land. That year, millions of acres burned out of control, and a haze of smoke covered the region, casting the landscape into shadow. Because the sun was blocked by the smoke, the forests and scrublands that were not directly destroyed by the fires suffered huge drops in a variety of crops, including many flowering and fruiting plants. This is where we get back to the fruit bats. With forests disappearing and fruiting and flowering plants deeply depressed in the smoky haze, the flying foxes were running out of options.

In the late 1980s, Malaysia (a Muslim country) had dramatically increased pig production to meet the needs of the Chinese in Southeast Asia, especially in Singapore, which was phasing out farming because it was running out of land. The Malaysians, being conscientious about land use efficiency, grew mango trees on their pig farms. In fact, mango production expanded at about the same rate as pig production in the 1980s and 1990s.

The Malaysians also let the pigs go into outside enclosures for fresh air. Guess what. The fruit bats found the fruit trees. They pooped into the open pig pens and dropped partially eaten fruits. The pigs, which will eat anything, ate everything; the virus came along for the ride. The viruses liked the pigs. Only a few pigs got sick, but the virus had a multiplicative heyday. The farmers picked up the virus from the pigs. Only a mass slaughter of pigs stopped the epidemic.

In several years following the epidemic, there were outbreaks of Nipah virus, clinically appearing as fever, headaches, and "altered levels of consciousness," in Bangladesh and West Bengal State in India. In these countries, people who climbed date palms and drank raw palm juice were getting the disease. The palm juice was collected in open-topped clay pots. Fruit bats, feeding in the same trees, had apparently dropped half-eaten bits of fruit, as well as, possibly, feces, into the pots. In this case the preventive measures were simpler and less traumatic than in Malaysia: put cloth covers on the pots.

SARS

At the end of May 2003, I was on a train from Ottawa to Toronto, returning from the Writers' Union of Canada annual general meeting. Outside the window, the countryside was greening into spring. Inside, it was all worry and fret. Several of my traveling companions wondered if they should be going to Toronto at all. A week after WHO had taken Toronto off an international travel advisory list for a newly identified disease called severe acute respiratory syndrome, a second epidemic wave was coughing its way across the city. More than two thousand people were in self-quarantine—which in Canada that year was a somewhat leaky concept, relying on goodwill and a sense of civic responsibility. Canada was unlikely to announce, the way the Chinese government did, that it would execute or jail for life anyone who broke quarantine.

In a democratic society, who out there might be spreading death and mayhem? Could we rely on a sense of public responsibility? Did my friends need to wear masks in the street? Would their lives be in danger? Remarkably, I reflected, despite my general cynicism about the self-centered consumerism of our culture, people did generally heed the call to quarantine themselves. When they broke quarantine, their activities seemed to be based on ignorance rather than bad will.

Who would have thought that two deaths in Toronto could instill such fear? Sui-chu Kwan, a seventy-eight-year-old woman, died on March 5 of an apparent heart attack. A week later her forty-four-year-old son, Chi Kwai Tse, waited eighteen hours in an emergency room before dying of what was apparently tuberculosis. Yet, as two of the non-Chinese casualties of what has been called the twenty-first century's first pandemic, as indicators of both the consequences our past (in)actions and of our probable future, the deaths of Mrs. Kwan and her son have as great a significance as 9/11. Those of us who work near the health system were astounded at the chaos, infighting, backbiting, and heroism. Despite massive mobilization of doctors, nurses, epidemiologists, politicians, and weary volunteers, more than four hundred Canadians fell ill, and forty-four died.

Worldwide, eight thousand people got sick and more than eight hundred died. But the death toll is not really the story. The story is how awkwardly the most sophisticated, scientifically based medical systems in the world responded to what surely should have been a routine event: someone stricken with a potentially fatal infectious disease. The story is about a kind of cultural amnesia and the cavalier satisfaction of the recently healthy.

The first cases of a kind of unusually severe and acute pneumonia had been reported from Guangdong Province in south China as early as November 2002. By February, cases were being reported in Vietnam. That month, Carlo Urbani, a physician and an officer at WHO, began documenting cases of this apparently new disease in a French hospital in Hanoi; by early March, he expressed concern that almost two dozen hospital workers had been infected. He called the new disease severe acute respiratory syndrome. "Syndrome" is the medical word used when you have a cluster of symptoms but you don't actually know if they reflect one disease, or, if they do, what that disease is.

By mid-March, cases had been reported from Hong Kong and Canada. High-tech laboratories around the world went into high gear to try to identify the agent, and WHO sent out a worldwide alert. Researchers had suggested that certain individuals might be "super-shedders"—that is, people who were spraying out extra-large numbers of viruses. Mrs. Kwan had returned to Toronto from a visit to Hong Kong, where she had stayed in the same hotel as a professor from China. The professor, unknown to himself, was a super-shedder.

WHO began to issue advisories warning travelers to avoid certain places if they could. By the third week of April, that list included not just cities in eastern Asia but also Toronto. By the time I was making my train trip to Toronto with my wary fellow writers, the travel advisory had been lifted, exhausted health workers were struggling to contain a new outburst of disease, and Dr. David Naylor, dean of the Faculty of Medicine at the University of Toronto, had agreed to lead a special advisory committee on SARS and public health.

Many of my colleagues have wondered how the Ontario health system had bungled things so badly and how the system might be fixed to prepare us for the next pandemic. Canadian researchers were able to sequence the genome of the offending virus, but it took more than a month before basic, low-tech public health measures could be applied to contain the disease.

Admittedly, gene sequencing is sexier than sharing data, developing clear lines of communication, and following hospital infection control procedures. I don't blame the gene sequencers. They were doing what they do best, and their work is important in tracing the origins of the disease and developing tests and vaccines. But, historically, diseases have not been controlled by gene sequencing; the most effective solutions are based on careful clinical observations and good old-fashioned health care, where "care" actually means care. Dr. Naylor's report on the Canadian response to the epidemic addressed many of those issues.

The advisory committee that Dr. Naylor headed, set up by Carolyn Bennett, the astute federal minister of state for public health, among other things resulted in the establishment of the Public Health Agency of Canada, an agency that has since done much to identify possible emerging disease threats and to sort out strategies for dealing with them.

Whatever the problems were once the epidemic started, I was more interested in where this pandemic came from. In my view, people showing up at hospital doors with SARS or avian influenza represent failures of public health, no matter how efficient or triumphant the medical response. In the long run, it seemed to me, the important questions had to do with identifying how these diseases emerge in the first place. So where and how did SARS emerge?

It seemed to come out of southern China in about 2002; that much was known. But why there? Why then?

In mid-March 2003, scientists working on the virus first suggested that it might be something related to measles and to animal diseases such as canine distemper and rinderpest in cattle, or at least in the same family, the paramyxoviruses. WHO officials nodded their heads, and investigators went off looking for these viruses, which, in many cases, they found.

Other scientists then proposed a coronavirus, which under an electron miscroscope is said to have a corona, or crown. To me these viruses look like spheres with pins stuck into them, but I don't know a lot about crowns. Some types of coronaviruses cause the common cold; others cause respiratory disease in chickens (infectious bursal disease); still others are associated with diarrhea in dogs. The scientists called it SARS-coronavirus, or SARS-COV. So the high-tech lab people had come through fast.

But what about the bigger questions about where the disease came from and hence how it might have been stopped or contained in the first place? Avian influenza had jumped from chickens to people in Hong Kong in 1997. Had SARS followed a similar path from animals to people?

Investigators familiar with the wide variety of animals farmed, caught, trapped, marketed, and eaten in south China, especially Guangdong Province, tested more than half a dozen types of animals from a market. They found evidence for the new virus in a raccoon dog and in several masked palm civets (*Paguma larvata*); although the latter are sometimes called civet cats, and they have a cat-like look, these fruit-eating, palm-dwelling animals are closer to mongooses.

Civets are gourmands of a sort. They love durian. Durian is a Southeast Asian fruit worth millions of dollars in trade and sales. When I lived in Java, a friend of mine who *liked* it declared it was like eating ice cream in an outhouse. The Indonesians wouldn't let people carry it on buses or airplanes because of its powerful odor but still swore by its wonderful taste.

Civets love to eat a particular kind of Vietnamese coffee bean. The beans travel through the civet's digestive tract and, when they emerge, voila, they are *caphe cut chon,* or fox-dung coffee, which some people prize. From about the tenth century BC until recent times, civets have also been prized for their powerful musk, which was used to stabilize perfumes. Some (maybe a hundred a year) were imported to United States as pets; importing them into North America is now, as a result of their association with the SARS virus, forbidden. The Chinese and Vietnamese prefer to eat them.

In any case, once the finger was pointed at civets, thousands of them were slaughtered. Still, there were lingering doubts that civets were the natural reservoir and not simply part of the epidemic, which emerged from some other source. Investigators persisted. Apart from a situation in which a waitress and a customer in Guangzhou, China, picked up SARS in the winter of 2003–04 from serving and eating civets near cages where the civets were kept before being eaten, the data were ambiguous. Could it be that civets were not the host reservoirs at all, but were victims of the disease, like crows were to West Nile virus, or, at most, launch pads or amplifiers of the invading viruses, which came from elsewhere? Investigators could not find any widespread infection in either farmed or wild civets. Where else could a person look?

Bats also are found hanging around (most unwillingly) in the markets of south China and Southeast Asia, where they are used for food and medicine. Given their history with Nipah and Hendra viruses, some scientists decided to check them out. All the bat species they tested had evidence of infection with SARS-COV-like viruses without being sick themselves. They were a perfect reservoir, silently infected, with no one watching them.

SARS-COV-2

After the alarm bells of avian influenza and SARS, one would think we might have been ready for COVID-19. If we were still living in the world of 2003, without such rich peripheral vision—instant access to many different global news sites and global connections through webs of social media—we could perhaps have been forgiven for not noticing the coming pandemic. On the other hand, with so many possibilities, it seems we can become paralyzed, like one of my backyard chickens frozen in the beam of my flashlight. Maybe evolution hasn't prepared us very well for the twenty-first century. Perhaps, after all, we can forgive ourselves for not noticing.

Most of the stories about the explosive emergence of SARS-COV-2 in late 2019 begin with a market in Wuhan, China. The implication seems to be that the pandemic's beginning was sparked by a mixture of bad luck and unscrupulous meat-sellers in a crowded market. This may be partly true. To get a more complete backstory, however, requires us to sharpen our collective peripheral vision; the tale of the birth of the pandemic of 2020 includes a catastrophic disease in pigs, bird flu, hundreds of millions of New Year's partygoers, and market vendors who may (or may not) have been unscrupulous, but were more likely just canny capitalist entrepreneurs. Humor me, then, and let me begin with the killer epidemic in pigs.

African swine fever virus causes a rapidly fatal hemorrhagic fever in domestic pigs. This modern, pig-killing version of the virus seems to have evolved, in Africa, about AD 1700, from a soft tick virus. In the ticks, as well as in wild bushpigs, the virus doesn't seem to cause any discomfort. At the end of the nineteenth century, after millions of cattle in Africa died from rinderpest, farming with white pigs imported from Europe and East Asia expanded rapidly, as did the range of the virus. Since then, ASF has appeared in various parts of the world, sometimes through the populations of wild boar imported by hunters and gourmet diners. Countries have tried to control the disease by slaughtering infected pigs. Some people may have heard about ASF as the punch line in a joke, when, in 2019, Denmark proposed building a wall along its southern border to keep out wild boars from Germany.

In 2018 and 2019, more than 200 million pigs in China died from—or were killed to "stamp out"—ASF. That was about half of the pigs in China and a quarter of all the pigs in the world. According to a November 2019 report in *Bloomberg News*, "Even if all the world's supply will be exported to China, it can only cover 10 million metric tons of the 22 million metric ton pork supply deficit." At the end of 2019 and in early 2020, hundreds of millions of Chinese people were out in the markets and malls searching for scarce meat to celebrate the end of the lunar Year of the Pig and launch of the Year of the Rat.

In January 2020, the *South China Morning Post* reported an outbreak of avian influenza in Hunan province, which is geographically adjacent to Hubei province. The outbreak killed more than half of the 7,850 birds on a single farm. The strain of bird flu involved was H5N1; tens of millions of birds have died from avian influenza since the early years of this century, and hundreds of millions have been slaughtered. Still, WHO has reported that, since 2003, avian influenza has infected fewer than a thousand people worldwide. About half of them died, and while each death is a personal tragedy, five hundred global deaths, mostly from people with "intimate" chicken contact (slaughtering, preparing), is not the global catastrophe that some predicted. The low numbers might reflect that we over-reacted to the threat, or that we were successful in our responses.

If pigs and chickens were in short supply, the Huanan Seafood Market in Wuhan, Hubei Province, China, was well-stocked with other species to help their customers prepare for the New Year celebrations. These included peacocks, wild rabbits, snakes, deer, crocodiles, turkeys, swans, kangaroos, squirrels, snails, foxes, pheasants, civets, ostriches, camels, cicadas, frogs, roosters, doves, centipedes, hedgehogs, and goats.

It should not have been a big surprise, then, given this perfect storm of animal disease and celebratory feasting, that a few people picked up a virus from one or more animals at the market and brought it home to share with friends and family. The disease they suffered ranged from dry, hacking coughs and fevers to shortness of breath, and, in a small percentage of people, death.

The new virus, rapidly detected and described by early 2020, was classified by virologists as a coronavirus, with its characteristic crown-like structure—the seventh known coronavirus known to infect people. Most diseases associated with coronavirus infections in people are mild, like the common cold. A few are more serious. In chickens, they cause infectious bronchitis, in cows and pigs, diarrhea; those strains don't normally infect people. As a veterinary epidemiologist, I really don't care if they wear crowns or jeans: How do they behave? Are they zoonoses—that is, transmitted from other animals to people? Do they adapt to humans and transmit from one of us to others, like H1N1 and SARS? Are they serious introverts, staying close to home, or extroverts, dancing easily around the world?

This newly recognized virus seemed to be related to the one associated with SARS, which is believed to have jumped from bats to civets (or some other animals) and hence into people. It is also a close cousin to MERS-COV, a coronavirus that emigrated from bats to camels, and from camels to people, resulting in a disease called MERS (Middle East respiratory syndrome).

Initial laboratory investigators suggested the novel coronavirus was genetically similar to a virus found in snakes; subsequent studies suggested that the viral genome was more closely related to a coronavirus found in pangolins, a type of scaly anteater. There has been no evidence that it spilled directly from bats to people. Pangolins, sort-of-famous for being Prince William's favorite animal, are a critically endangered species. Just as the international research related to avian influenza uncovered global webs of illegal trade in wild birds, the COVID-19 pandemic has begun to pull back the curtain on pangolin smuggling. With its meat desired by some wealthy people as a sign of prestige, and its scales sought by desperately gullible people for their alleged medicinal value (like fingernails and hair, they are made of ordinary keratin), pangolins had, by early 2020, become the most widely trafficked wild animal in the world.

Also by early 2020, it looked as if this virus—later dubbed by scientists SARS-COV-2—was sufficiently well-adapted to people that it was being transmitted directly from human to human. Like SARS-COV, the new virus is transmitted mainly through respiratory droplets produced when infected people cough, sneeze, or talk. This route is the main reason for the public health directives on social distancing put in place nearly everywhere around the world by March of 2020.

There is some laboratory evidence suggesting that the virus is stable for a few hours on cardboard, and a few days on plastic and stainless steel, as well as being present in feces and urine. As with SARS-COV, this kind of transmission through inanimate materials (called fomites) is possible, but (as of April 2020) is not believed to be common.

Not surprisingly, economic uncertainty followed closely on the heels of the pandemic. What also should not have come as a surprise was the fierce battle over the narratives about how the virus emerged, how it spread, and especially how the rapidly growing pandemic would be managed and stopped. With control of the narrative comes political and economic profits and power. Is this a failure-of-liberal-free-trade story? A failure of the predict, command, and control story so beloved of autocrats and some scientists? A way for pharmaceutical companies to distract the public from the opioid crisis and shift their attention to another possible profit-maker? A story of nefarious free marketeers? Of strutting rich boys playing food-trading games as if they were dealing with baseball cards and bubblegum? The emergence of unexpected consequences of mostly well-intentioned people working by the rules in a dysfunctional agri-food system? Or a story of complex human behaviors embedded in a web of largely unexamined ecological webs? Or, as some of us believe, are all of these stories true, and, if so, what can we do?

(II)

RATS, BATS, AND MONKEYS: LASSA, EBOLA, AND MARBURG

HEMORRHAGIC FEVERS: the name says it all, and it conjures up many of our darkest fears. These diseases all, it seems, come out of the rain forests of Africa. They are described as "terrorizing" or "stalking" the human race. The agents are always "deadly," and they are "killers." They are reason enough for firebombing African villages and threatening to do the same to American cities, as in the movie *Outbreak*. They have names such as Lassa, Ebola, and Marburg. Marburg? Germany? Yes, and the strange thing is that if you are a virologist, used to looking at electron micrographs, you would say that Lassa was the one that doesn't belong. It's an arenavirus, you would say. The others are filoviruses; they look kind of stringy.

For non-virologists, what is more important is this: they all, more or less, start with malaise and aching muscles and then move on to high fever, coughing, diarrhea, pain everywhere, and, in the worst cases, bleeding all over, brain problems, shock, and death.

In 1969, a nurse in Lassa, North-Eastern State, Nigeria, came down with fever, aches, and pains—the unremarkable symptoms of serious tropical diseases such as malaria, or diseases of poverty such as typhoid fever. When she didn't get better after being treated locally, she was flown to a hospital in the Nigerian city of Jos. She died. One of the nurses who had taken care of her there also died. A third nurse, who had helped with the post-mortem of the first, became ill and was flown home to the United States in a commercial airliner. The Yale Arbovirus Research Unit isolated a virus from her blood and called it Lassa virus. A laboratory worker at the Yale laboratory fell ill, was treated with a transfusion of serum from the nurse, who was recovering and presumably had antibodies. The lab worker recovered. Five months later, another laboratory technician at the lab died of Lassa fever. No one knows how it happened.

The natural home of the Lassa fever virus is an African soft-furred rat, also called a multimammate rodent, *Mastomys natalensis*. *Mastomys* are at home in savanna woodlands common throughout parts of sub-Saharan western Africa. The rodents are sometimes described as pests and sometimes as food. I guess that depends on where they are found and how hungry you are.

Lassa fever virus belongs to a larger family of rodent-loving viruses. Members of this family include Machupo virus (causing Bolivian hemorrhagic fever), Junin virus (Argentinian hemorrhagic fever), Guanarito virus (Venezuelan hemorrhagic fever), and lymphocytic choriomeningitis virus (found in Europe and North America and causing "flu-like" lymphocytic choriomeningitis disease, or LCM, and only rarely meningitis in people).

In the early 1970s, Lassa was considered a rare disease, which killed about half the people it infected, mostly in hospitals. But then, many diseases are considered serious when first discovered, simply because the most serious cases are the ones first reported. For a few years, small outbreaks with high death rates were reported from hospitals in Nigeria and Liberia. Then, investigators of a hospital outbreak in Sierra Leone reported that less than 10 percent of the people infected there picked it up in the hospital. The case-fatality rate was less than 5 percent.

Using this news as a starting point, a group of investigators from the CDC, led by Karl Johnson and, later, Joseph McCormick, set out to describe the epidemiology and ecology of Lassa fever virus in its natural setting. They found that the disease was endemic in much of western Africa. A few hundred thousand people a year may get infected, and 15 percent of them die with "multisystem" failure—which basically means that all their body organs shut down.

Occasionally, people still pick up Lassa fever virus in Africa and carry it home to other parts of the world. In 2004, a businessman returned to New Jersey after a five-month stay in Liberia. Toward the end of his visit to Africa, he developed fever, chills, sore throat, diarrhea, and back pain. Doctors in the hospital in Trenton, New Jersey, treated him with antimalarial drugs and antibiotics (for typhoid fever). He got worse and had trouble breathing. The doctors went down their list of possibilities and came up with yellow fever and Lassa fever as candidates. Both of these require expensive antiviral treatments. He died before they could start treatment. Since getting sick, he and/or his bodily fluids (blood) had come into contact with fellow airplane passengers, train passengers, family members, and laboratory workers. Nobody else, fortunately, got sick. However, as with Ebola, Lassa fever is serious, but the virus is not easily transmitted, requiring close contact with infected fluids.

In 1967, twenty-five laboratory personnel in Frankfurt and Marburg, Germany, and in Belgrade, Yugoslavia, fell acutely ill with high fevers, muscle and joint pain, vomiting, diarrhea, rashes, and bleeding, both internally and from the nose. Some of the people who took care of them also got sick. One spouse was infected through her husband's semen. Seven of the infected people died.

All of those originally infected were working with tissues and blood from African green monkeys (*Cercopithecus aethiops*) imported from Uganda. Cell lines derived from African green monkeys are used for vaccine research and various kinds of testing. For instance, the disease associated with the infamous *E. coli* O157:H7, sometimes called hamburger disease, is also called verotoxin-producing *E. coli*. The actual disease is caused by a so-called verotoxin, which means that it kills African green monkey kidney cells in the laboratory. The original monkeys had not appeared to be ill, but there had been daily deaths among the monkeys in transit from Uganda, and experimentally inoculated monkeys of various sorts did become ill and die.

For a variety of good reasons, initial suspicions fell on the monkeys as a reservoir host. Non-human primates are particularly dangerous sources of disease for people. As we have been reminded many times since Darwin identified non-human primates as our genetic relatives, the infectious-disease jump from a monkey to a human being is more of a small step than a great leap.

For instance, a form of simian malaria, caused by the parasite *Plasmodium knowlesi*, is widespread in both people and Old World monkeys in Southeast Asia, complicating the already huge global challenge of dealing with malaria.

Perhaps the most tragic example of transmission of diseases from non-human primates to people is that of HIV. Transmission of simian immunodeficiency virus (SIV) from one subspecies of chimpanzee (*Pan troglodytes troglodytes*) to hunters (through cuts) and its transformation into a global pandemic is a classic case of multiple unintended consequences. In the 1920s, health authorities in French-administered African territories initiated large-scale programs to reduce the burden of sleeping sickness. The only available treatment was an injectable arsenic-based drug; there was a shortage of needles and syringes, no easy way to sterilize equipment, and, in any case, no understanding of how viruses could be transmitted through contaminated injection equipment. From there, the human-adapted HIV spread into adjacent French- and Belgian-administered territories through both medical injections and sex workers (who provided support services for colonial industrial development). In the early 1960s, Haiti contributed more than a thousand technical workers and teachers to help rebuild the Congo after the Belgians left. They returned to Haiti at a time when their country was actively exporting corpses and blood plasma to American medical schools.

The rest of the tragic story of HIV/AIDS has been well recounted in both nonfiction and fiction and I shall not belabor it here.

My point is that there were good reasons for researchers, when faced with Marburg disease, to first look at monkeys. The monkeys in Marburg had been imported from an island in Lake Kyoga, a shallow, swampy lake near Mount Elgon, on the Kenya-Uganda border, where I had been part of a large project investigating zoonotic sleeping sickness. The fact that many of the monkeys infected with Marburg virus died of the infection should have been a clue that they were not, in fact, the reservoir. A host reservoir is usually an animal that can harbor the agent without getting sick.

Although people can get the virus, and the disease, from capturing and eating sick monkeys, we are now fairly certain that the natural reservoir host is the Egyptian fruit bat, *Rousettus aegyptiacus*, a highly social cave dweller that across large parts of Africa is an important pollinator and seed disperser.

In the twenty years after the initial outbreak, only half a dozen incidents of Marburg virus disease were recorded, in Kenya, South Africa, and Zimbabwe, involving altogether a few dozen people, about a quarter of whom died. This changed abruptly in 1998, when an outbreak of Marburg disease started among "unofficial" gold mine workers in the Democratic Republic of the Congo; 149 people were struck with the disease and

123 of them died. Then, in 2004 and 2005, an outbreak of fever, vomiting, coughing, hemorrhaging, and diarrhea was reported from northern Angola. One hundred and twenty-four people got sick; three-quarters of them were children under five years. Of those who were sick, 117 died.

Since then, there have been small, sporadic outbreaks of Marburg virus disease, or single cases, often of people who visited bat caves, but no major epidemics.

ALMOST A DECADE after the 1967 Marburg disease outbreak, dramatic epidemics of a hemorrhagic disease struck southern Sudan and northern Zaire. In the Sudan, about three hundred people got sick and more than half of them died; in Zaire (now the Democratic Republic of the Congo), about three hundred people got sick, and about 80 percent of them died. The virus that was isolated from the victims looked exactly like the Marburg virus, but it turned out to be something different. Doctors called it Ebola virus, after the Ebola River, one of the headwaters of the Congo (Zaire) River.

Since those 1976 epidemics, scientists have determined that there are at least four different strains of the virus. The Maridi strain, which has struck in southern Sudan and northwest Uganda, kills from a third to two-thirds of its victims. The Zaire subtype, involved in attacks in Zaire, Gabon, the Democratic Republic of the Congo and the Republic of Congo, kills from 60 to 90 percent of its victims. The third African strain, from the Ivory Coast, is known to kill chimpanzees but not people. Another strain of Ebola, the Reston virus, was isolated from monkeys at a quarantine station in Reston, Virginia; the cynomolgus monkeys (and the virus) came from the Philippines, and, although it has been capable of killing the monkeys, inspiring writers of best-selling paperbacks, and instilling panic in the general public, this particular strain does not appear to cause serious disease in people.

Since the 1970s, the frequency of serious outbreaks of Ebola virus has been increasing in Africa. Between 1976 and 2013, WHO reported twenty-four outbreaks, involving 2,387 people; two-thirds of them died.

Then, beginning in late 2013, and extending into 2016, almost 29,000 people came down with the disease, 40 percent of whom died. The index case (the first reported case) of this epidemic was an eighteen-month-old boy in Guinea, who is thought to have picked it up from a bat. From that poor, rural village, the disease spread quickly to neighboring Liberia and Sierra Leone.

Once a person is infected, Ebola virus is spread by close contact between the sick individual and his or her caregivers, as well as through burial rites and the preparation of bodies for those burials. Controlling the spread of the disease thus requires culturally sensitive engagement with local people; caring for loved ones, and burial rites, are profoundly rooted in history and culture and are not changed through lecturing from medical experts. Often the very techniques that are so effective for treating disease (chain of command, control, reliance on technical expertise) are the least effective for promoting prevention and long-term health.

This Ebola outbreak stirred up panic in many Western countries. Unlike for SARS or COVID-19, the panic was not related to uncertainty. Medical authorities were certain that the virus, like the Marburg virus, was difficult to transmit. Some of the negative responses were clearly based on a combination of ignorance, racism, and xenophobia. In the U.S., these responses ranged from blaming immigrants from Mexico for the disease to schools closing their doors to teachers who had visited anywhere in Africa. Morocco, nowhere near the Ebola epidemic, canceled the Africa Cup soccer match, and, in an eerie omen of things to come with COVID-19, Mexico and Belize refused to allow a cruise ship to dock because a healthy passenger had handled Ebola-related lab specimens in a Dallas hospital.

To their credit, the CDC, after assuring Americans that they couldn't get Ebola from their Thanksgiving turkey, sent a team to treat patients at the center of the epidemic. They also trained almost 25,000 health care workers in West Africa and expanded laboratory capacity to test for Ebola virus in Guinea, Liberia, and Sierra Leone. Other non-African countries also stepped into the technical breach.

The National Microbiology Laboratory (NML) in Winnipeg was first opened in the late 1990s. In 2004, the lab became part of the Public Health Agency of Canada, which had been created in response to the Naylor report on how SARS was mishandled. The NML is the only Containment Level 4 (Biosafety Level 4) facility in Canada. Being one of the few laboratories around the world that works on both pathogens of people and other animals, it is ideally situated to work on viruses that cause Ebola disease, SARS, and the like. A team at the lab, led by Gary Kobinger, had developed an Ebola vaccine, but by 2014 had been unable to get WHO or commercial companies to take an interest in it. In August of 2014, Canada donated the vaccine to WHO. That November, a joint venture between two commercial companies bought rights to the vaccine, and started commercial production and field trials. By December of 2016, they determined that vaccination with rVSV-ZEBOV (the vaccine's technical name) provided between 95 and 100 percent protection against Ebola virus disease.

Index cases for Ebola outbreaks have not necessarily been people who handled bats; more often, they have reported contact with dead gorillas, chimpanzees, and duikers. These animals are important sources of what is called bush meat; that is, wild-caught meat. Non-human primates are particularly dangerous, because for a microbe, moving to a new home in people doesn't require major adaptations.

The underlying causes of these epidemics, including why people eat bush meat, are rarely addressed, at least with any depth or persistence. It's complicated, but not rocket science; difficult science, but not necessarily, in the conventional sense, "hard" science. In many poor countries, such as those of West Africa, there has been a perceived need, often encouraged by international businesses and donor agencies, for foreign exchange. At one time this was perhaps viewed as a way to bring in money to build up local industry and infrastructure. That fig leaf is long gone. The beneficiaries of this wealth generation are primarily in North America and Europe (and, increasingly, China).

What do these countries have that wealthy countries want? Timber and precious minerals seem to be high on the list. What follows is rapid, unregulated deforestation and mining, which can cause wars, which then lead to further deregulation. Liberia suffered back-to-back civil wars from 1989–97 and 1999–2003, the second initiated by a Guinea-backed group. In 2014, the République de Guinée reported anticipating US$50 billion of mining investments over the next decade. After the cessation of war, more than half of Liberia's forests were sold off to industrial loggers.

Poor and marginalized peoples are attracted to these money-generating activities (money for whom? we should ask), which then result in these populations scrounging for food as best they can.

People eat particular foods for a wide variety of reasons. Some of these—such as avoiding pork—are related to cultural or religious identity. For others, the preferred food relates to availability. I grew up with borscht made from cabbage, carrots, and potatoes, foods that stored well over the long winters where my ancestors came from. I have an Acadienne colleague who declares that the traditional (therefore the best) tourtière is made with wild game hunted by family members.

In the parts of Africa where Ebola appeared, some people hunted for bush meat not only because they liked it, but because they were away from the subsistence farms that had been their homes. Hunting bush meat increased food security in their new environments. The easiest meats to get in the areas newly cleared and invaded by foreign industries were from animals already debilitated from diseases such as Ebola, or from hunger associated with loss of habitat. As people probe or are pushed deeper into new ecosystems in search of food, the bush meat trade has become a continuing potential source of new diseases, including, probably, several strains of the viruses that cause AIDS. The human epidemics of Ebola virus disease appear to have been preceded by, or associated with, deadly epidemics in animals, and biologists are afraid that Ebola may be devastating some wildlife populations in central Africa.

The usual response? Wealthy donors rush in and lecture the local people about eating bush meat, even as wealthy countries and wealthy classes within countries are the economic beneficiaries of their disease. Zoonoses are, to a point, good for (some) businesses.

Although Egyptian fruit bats don't travel out of Africa, people move a wide range of species, legally and illegally, out of Africa every day. The animals are transported to become pets, to be used in research, or for food, or just because that's what humans do. In 2003, monkeypox (fortunately not a fatal disease in people) got into the United States when an animal distributor imported a variety of rodents from Ghana. These included species such as rope squirrels (*Funisciurus* sp.), tree squirrels (*Heliosciurus* sp.), Gambian giant rats (*Cricetomys* sp.), brushtail porcupines (*Atherurus* sp.), dormice (*Graphiurus* sp.), and striped mice (*Hybomys* sp.). The animals were then shipped to pet stores across the country and housed in the same stores that sold prairie dogs, those pesky things that farmers like to shoot. Evidently, there is a market for these little beasties as pets; apparently, the market does not know that these animals act as vessels for microbes out to see and colonize the wide world.

As of this writing, an epidemic of Ebola in the Democratic Republic of the Congo, declared in June 2019 to be a "world health emergency," is ongoing. Few in non-African countries are paying attention.

It may be time to re-examine some of the things we "just do."

(12)

SPLASHING THROUGH RAT PISS: LEPTOSPIRES AND HANTAVIRUSES

SWARMS OF rodents and warm, wet weather are a problem not just for the plague and Lyme disease but also for another group, or clade, as some microbiologists refer to it, of spirochetes, the leptospires. Leptospirosis, the disease, is caused by more than 250 species of spirochetes and may be the most common zoonosis in the world. Actually, we don't know how common it is, as reporting requirements and compliance are variable. The disease is most often reported in the Caribbean and Latin America, the Indian subcontinent, Southeast Asia, and Oceania. But not always.

In mid-June 1998, more than eight hundred triathletes lined up on the shore of Lake Springfield, Illinois, for the sixteenth annual Springfield Ironhorse Triathlon. These were healthy young men and women, ready to swim 1.5 miles in the clean midwestern lake, bike 45 miles across the flat landscape, and then run another 10 miles. Swimming unseen alongside them in the water were little schools of leptospiral bacteria, which insinuated their way undetected across the mucosal surfaces of the eyes, mouths, noses, vaginas, and penises and into the bloodstreams of these athletes.

For the next couple of weeks, the tiny speedster bacteria dashed around the blood vessels until they got stuck in a variety of small capillaries, especially in the kidneys, where they stirred up trouble, causing local damage and bleeding. A week to two weeks later, dozens of these prime young specimens were suffering from fever, chills, headache, muscle pain, diarrhea, and red eyes. Twenty-three were hospitalized. Nobody told them that rats or cows or some other animals had been peeing into the lake or that rats carry not only plague or hantaviruses but also, often chronically, in their kidneys, leptospires that they ceaselessly piss out into the environment. Nobody told them that leptospires love to hang out in fresh water, damp soil, mud, and vegetation. Maybe they thought that the white-water rafters who picked up the same disease in Costa Rica in 1996 were somehow different. That was Central America, after all.

Given a few economically privileged adventurers complaining of muscle pains and fevers, one might be forgiven for not lifting this disease to the top of the global agenda. None of them, after all, came down with the most severe form of the disease, which results in kidney failure, bleeding from the lungs, and jaundice.

This severe form is called Weil's disease, not after the French mystic Simone Weil, although her passion for working with the poor would suit those populations who suffer most from Weil's. The disease was named after Adolf Weil, a nineteenth-century physician, who described it but didn't actually discover it. Japanese researchers early in the twentieth century discovered the spiral organisms with hooked ends in the kidneys of coal miners, among whom the disease was common, and also showed that rats were the primary carriers. Much of the early work was carried out in the warm, moist environments of South and eastern Asia.

A few queasy triathletes did not succeed in putting the disease very high up on the U.S. worry list. Most of the world, however, does not live in North America. Dairy farm workers in New

Zealand get peed on by infected cows, and throughout the tropics millions of workers in rice fields, cane fields, or mines wade through, and occasionally accidentally imbibe, the pee of rats or other small mammals. Although rats are considered the chronic carriers (in their kidneys), dogs, cows, and pigs can also carry and shed leptospires. Livestock or dogs can be vaccinated, but the vaccine prevents disease; it doesn't prevent the organisms from snuggling up in the kidneys, and it doesn't prevent the little sportster spirochetes from shooting the yellow rapids out the urinary tract. Most of the time the organisms infect the kidneys, damage a few small blood vessels so that the blood leaks out, and cause a bit of fever but don't do anything more serious than that.

What is more worrisome—if you are looking for something to distract you from the latest pandemic or from wondering where the kids are tonight—is this: in 1995, after Hurricane Mitch flooded Nicaragua, two thousand people fell sick and more than a dozen died. They *did* bleed from the lungs. Between 1996 and 1998—just before and during a major El Niño–Southern Oscillation (ENSO) event—there were serious outbreaks in Ecuador, Peru, and Brazil.

In 1996, in one hospital in Salvador, Brazil, several hundred people got the disease, of whom fifty died. The best predictors of whether patients would die were "altered mental states," "respiratory insufficiency" (that is, they couldn't breathe), and kidney failure. That same year, over just two months (February and March), fifteen hundred cases were reported in Rio de Janeiro. Not surprisingly the most important predictors of getting sick had to do with living in steep areas where there was poor sewage management, poor garbage collection, and no distribution of clean water. Studies in Peru found that living in slum conditions and not wearing shoes in the field put you at risk of getting sick. Go figure. It seems somewhat pointless to post an advisory, as the CDC did for Lake Springfield, to stay away from the water.

Attending large recreational events and living in slums are the two biggest predictors of where the disease is likely to hit. I suspect these two do not include the same people. And which activity do you suppose is likely to have more resources devoted to solving its problem?

Leptospirosis appears to be increasing in dogs in many parts of the world, probably, in part, because of global warming. Furthermore, there is shifting in the types of leptospirosis that are infecting animals. The strains vaccinated against are less likely to multiply and spill into the environment. As is happening with avian influenza, vaccination is inadvertently selecting for different strains. The vaccines don't cross-protect. The bacteria breed and evolve a lot faster than we and our much-vaunted biomedical technologies are capable of. And when sudden, heavy rainfalls occur as part of climate change, the rats piss and run (urinate and migrate, for the non-veterinarians who are reading this). Where they run to is where billions of people in the world now live—large urban slums.

WHILE WE ARE on the subject of problems associated with rodents, it is worth mentioning hantaviruses. This family of viruses is spread from small, wild rodents to people when the animals' saliva, feces, and urine are aerosolized, and people breathe it in with dust. Not a pretty thought.

The Four Corners area of the southwestern United States is a Navajo Tribal Lands area on a high plateau where Utah, Colorado, New Mexico, and Arizona meet. In the spring of 1993, people began wandering into local medical centers with what at first appeared to be common flu-like symptoms: fever, muscle aches, coughing. The only trouble was that after the first day or so, the illness it didn't behave like the flu. Of the first dozen or so who showed up at the medical centers, about three-quarters died rapidly from respiratory failure. In the week after May 14, near

Gallup, New Mexico, a young athlete, his fiancée, and four other people died. Over that summer, dozens more people came down with this new disease, and the case-fatality rate dropped, but not dramatically enough to make everyone relax.

The CDC did what it does best. It mounted a concerted effort to isolate and identify the infectious agent, right down to its molecular structure. The people at CDC may sometimes be arrogant and have trouble dealing with a lot of ecological and social determinants of disease, but they are *really* good at this technical stuff. They were able to identify a previously unidentified hantavirus, which was named, appropriately, *sin nombre* ("no-name"). The disease was called hantavirus pulmonary syndrome.

Since 1993, scientists have learned a lot about this family of viruses. In the early 1950s, during the Korean War, more than three thousand United Nations troops came down with a "new" disease that authorities called Korean hemorrhagic fever. No one really knew what was causing the acute fever, bleeding, and kidney failure; 10 to 15 percent of those affected died. That version of the disease is now called hemorrhagic fever with renal syndrome, which is somewhat more clinically (but somewhat less historically) informative. More than a hundred thousand people suffer from it every year in China alone. The actual virus wasn't identified until 1976, when Korean investigator Ho-Wang Lee isolated it and named it Hantaan virus, after a river in the demilitarized zone, where the disease is common. The medical diagnosis in the troops was made more than forty years after they got sick, based on stored serum samples and medical records. It is one of those cases that I use to justify not throwing out all those boxes of stuff that might be useful some day.

Variations on this disease, caused by a whole family of similar viruses, have been reported from places as distant from each other as Scandinavia and Manchuria, going back to the early part of the twentieth century. In Scandinavia, the disease seemed less serious than the Asian form and was called nephropathia endemica, which means it is a disease that is always around and affects the kidneys. Another form, similar to that in Asia, occurs in the Balkans. Some investigators have suggested that viruses from the same family may contribute to chronic hypertension and kidney disease in the United States.

At about the time the 1993 outbreak was pumping adrenaline through the bodies of medical investigators, James LeDuc and several colleagues from WHO summarized much of what was known about hantaviruses to that point. They suggested that the viruses "are very widely distributed and likely to be responsible for human disease in many parts of the world where they are not now recognized." Since that was written, an increasing number of cases and outbreaks have been recognized in North and South America, and a great variety of both viruses and rodent reservoirs have been identified. Researchers have also done detailed investigations into the lives of small rodents, their diseases, and human relationships to them. Apart from the Norway rat, which can carry the disease, other carriers include the deer mouse, white-footed mouse, bandicoot rat, meadow vole, reed vole, musk shrew, European common vole, yellow-neck mouse, small-eared pygmy rice rat, Mexican harvest mouse, cotton rat, prairie vole, California vole, vesper mouse, long-tailed pygmy rice rat, and Siberian lemming.

In "To a Mouse, on Turning Up Her Nest with the Plough," Robbie Burns lamented (as far as I can tell, struggling with my devil's tongue English understanding of the Scot's mind) disturbing the home of the little field mouse and the rupture with nature that her terror signified. "Wee, sleekit, cowrin, tim'rous beastie, / O, what a panic's in thy breastie!" he wrote to his "fellow-mortal" and "earth-born companion."

Had Burns known about hantavirus pulmonary syndrome, he might well have panicked himself—and our language and culture would be the poorer for it. Many cases in North America occur when people clean out cottages in the spring, disturbing nests and sending up clouds of mouse excretions.

Between 1989 and 2015, 109 cases and 27 deaths were reported in Canada. In the U.S., 728 cases and 262 deaths were reported between 1993 and 2017. Although disturbing the nests of rodents explains many individual cases, it doesn't usually explain outbreaks. In these instances, some combination of human behavior and ecological change is usually at work.

To understand these, we have to cast a wider net. In the several years before 1992, the Four Corners plateau was its usual semi-arid self. That year, as part of the ENSO, heavy rains fell. These rains led to a big crop of piña nuts. A big crop of piña nuts meant a much better survival rate and population boom for field mice. The spring of 1993 was an especially mousy year on the high plateau. The *sin nombre* virus (like all the hantaviruses) circulates in the rodents without causing much damage. It goes to the kidneys and lungs, and the little rodents appear to shed the virus indefinitely. The extra rodents invaded the cabins and cottages. More excreta to clear out. More contact with people. More cases.

Later, as if anticipating the medical-scientific obsession with predictive planning and control, worried maybe that people will try to control the weather, Burns cautions that "The best-laid schemes o' mice an' men / Gang aft agley." In trying to control everything, we have to come face to face with ourselves.

The message of hantavirus pulmonary syndrome for North Americans is that we should wear face masks while cleaning out the cottage and be especially careful after a mild, rainy winter and spring. Yet there is another, larger message that the rodents have for us, expressed in a language that has yet to be translated into those of politicians and land developers. How, this message might begin, do humans come into contact with all these rodents? People are driven to clear marginal land, and they disturb the "cowrin" beasties. They crowd into slums, and the rodents come to live with them. Perhaps we are seeing more diseases like leptospirosis and hantavirus pulmonary syndrome because we are looking harder for them. But I suspect there is more to it, having to do with poverty, inequity, and reckless use of the planet whose care has been bequeathed to us.

If wealthy people can sometimes protect themselves from ills carried by rodents, I confess it gives me some small satisfaction to see how the subtle voices of nature manage to slip past our expensive and sophisticated technico-medical defenses and catch our notice. Rodent excreta in the cottage. Rodent piss in the lake. "We are in the air," they whisper. "We are in the water. Welcome to our world."

(13)

ALL THE RAGE

IN FRENCH, rabies is called *la rage,* which is appropriate. Rabies draws on seemingly inexhaustible wells of fear and anger. The reasons for this go to a deep human anxiety about a dark wildness at the heart of nature. The wild beast is out there, just waiting for us, and the images of slavering dogs and their wild cousins, the wolves, are the face of that beast. Of the tens of thousands of people who succumb annually to this terrifying disease—sixty thousand reported in 2018—most are in the poor countries of the Southern Hemisphere, probably a quarter of them in India and China. In China, four epidemic waves, occurring at ten-year intervals from the mid-1950s on, killed at least 100,000 people. In the most severe of these, in the 1990s, more than 55,000 people are known to have died. After peaking at about 2,500 cases in 2004, the number of people who died from rabies in China dropped, reaching a low of 426 in 2018. In all countries, most human victims get the disease through dog bites.

The story of rabies is more complicated now than it was even fifty years ago. It used to be considered a disease of "mad dogs" that then bit people, and historically, that was mainly true. In much of the world, it still is. By the beginning of this millennium, other narratives about rabies had been constructed, based on a more complete scientific understanding.

After World War II, an epidemic of fox rabies swept out of the Soviet Union and westward across Europe. In the 1950s, fox rabies invaded southern Canada from the north, where it settled comfortably into the fox and skunk population.

Raccoon rabies was first recognized in Florida in the 1950s. From an initial small area, the disease spread north and west at about 25 miles per year. Then, in the 1970s, raccoon rabies was given a big boost; hunters from West Virginia decided they needed more raccoons to shoot and so brought several truckloads of them north. At least one raccoon in the bunch had rabies. By the 1980s and 1990s, raccoon rabies was speeding north through towns, cities, attics, and barns. By 1999, the rabid raccoons had reached the St. Lawrence River and were making their first crossings into Canada, just as Ontario had brought fox rabies under control through a wildlife vaccination program. There are two to three hundred raccoons per square mile in parts of southern Ontario: a disaster waiting to happen.

Rabies is kept alive by skunks in the Great Plains, coyotes in Texas, vampire bats in Latin America, and insectivorous bats just about everywhere. In the past decades, not only have scientists identified a much wider range of animals that harbor rabies than once was thought, but they have also identified different strains of virus. Some types of rabies virus are better adapted to raccoons, others to foxes, and still others to bats. In the mid-twentieth century, rabies was thought to be basically one disease, although a few oddball rabies-like viruses had been identified in Africa with names such as Oulou Fato (causing a non-fatal disease in dogs) and Duvenhage virus (which caused something that looked exactly like rabies but apparently wasn't). Fifty years later, the scientific literature was carrying reports on "rabies-like" lyssaviruses discovered in bats in Australia, Southeast Asia, Europe, and a variety of other unexpected places. The closer one looks, the more complex the picture seems and the less stable and homogeneous the virus appears.

The word "rabies" itself comes from the Sanskrit *rabhas,* 3000 BC, and means "to do violence." Chinese records from 555 BC use the same word for "mad dog" and the human disease hydrophobia. The Greek word *lyssa* is derived from a root meaning "violent"; according to Gerald Moss writing on mental disorder in antiquity (see Brothwell and Sandison, *Diseases in Antiquity*), *lyssa* originally seems to have meant "the rage that possesses the hero," but, in the *Iliad, lyssa* is connected with "dog," as in "mad dog."

So the connection between mad dogs and clinical rabies, often referred to as hydrophobia, is very old. Descriptions of rabies, connecting mental derangement, fear of water (hydrophobia), and the association with mad dogs go back to the twenty-third century BC, in Babylon. Three hundred and fifty years before Louis Pasteur, Girolamo Fracastoro, an Italian scholar, described both the disease and the means of transmission.

"Lyssaviruses" is the name now given to the group of viruses to which rabies virus belongs. Lyssaviruses are members of rhabdoviruses, which are rod shaped, with one end flattened and the other rounded, like a bullet. This may be an appropriate shape, given the seriousness of the disease.

The viruses, once insinuated through the skin into the muscle, multiply, find a nerve, and then float up the axon toward the brain, periodically pausing at junctions of nerve cells to reproduce. This trip to the brain may take anywhere from a few weeks to more than a year. Once in the brain, the viruses multiply again and then travel out the nerves to all the body's organs, including the salivary glands. With its brain commandeered and damaged by the viruses, the animal behaves in strange ways.

There are two main forms of the disease: furious rabies, in which the animal attacks unsuspecting victims; and dumb rabies, in which the animal mopes around and sinks into depression. An interesting study using game theory calculated that the best strategy for viruses to perpetuate themselves in animal populations would be to have variable incubation periods, variable symptoms, and variable social behaviors. It is pretty hard to learn which animals to avoid if they exhibit no consistent patterns. Natural selection seems to have provided for just such a wild-card combination.

Although descriptions of the disease, and its connection with badly behaving dogs, have been around for a long time, describing this devilish disease clinically did nothing to make it go away or even to tame it. Historically, some people killed themselves when they were bitten by what they thought was a rabid dog, more terrified of the disease than of any possible consequences in an afterlife. Other treatments have included holding people under water until they got better or drowned, a case of trying to treat the symptoms (fear of water) without getting at the cause, or feeding them various parts of chickens or dogs. No treatments, including the best technologies at our disposal today, are effective once the clinical signs set in.

Since the prehistoric human emigration from the gardens of our African and Middle Eastern Edens, dogs have been our most constant and faithful friends and companions. When we were hungry hunters, they guided us to prey. They have guided and guarded our sheep and cattle, sometimes paying for it with their lives. When we settled into agriculturally based settlements, they guarded our properties, barking at the approach of strangers. Our collective lives seem to have been a long march to the ultimate control of nature, and dogs have been a happy part of that march.

Control of nature has brought us many benefits. We grow genetically improved maize and coffee in neat, fertilized fields, accompanied by our tail-wagging canine friends, whom we have taught obedience. But nature has dark secrets. Just when we think we have control, it slips away. Volcanoes that seem to be asleep erupt in a grand, destructive passion. The microbial populations carried by the chickens destined for "every pot" rebelled. The dogs were more direct; they turned on us, snarling and biting and bringing death and madness; it is as if our own children were attacking us. Thus, when Louis Pasteur, in the last great scientific effort of his life, tackled rabies, he was facing the demons that always haunt us: the ones that say, what if all the things we think are normal and good suddenly turn on us?

Pasteur didn't need to stare down one of the most fearsome zoonotic diseases in history. He had figured out why beer, wine, and milk spoiled and how to prevent it: pasteurization. He had saved the French silk industry from death by a disease that ravaged the silkworms. He had, quite spectacularly, discovered and publicly demonstrated how to prevent anthrax.

In May 1882, at the farm town of Pouilly le Fort just south of Paris, before the watchful eyes of a skeptical media, Pasteur injected fifty sheep with virulent (disease-causing) anthrax bacteria. Half the sheep had been vaccinated with an attenuated form of the bacteria he had grown in the laboratory; the other half had been left alone. The twenty-five that were vaccinated lived. Those who were not vaccinated died. Some estimates are that the vaccination of sheep and cattle that followed saved the French enough money to cover their war reparations to Prussia for having lost the 1870 Franco-Prussian War.

Having a vaccine is great, but it is not a panacea. When a disease occurs sporadically, or if people are very poor, vaccination isn't practiced, which gives the disease a chance to make periodic comebacks. Anthrax is still with us, partly as a bio-terrorist threat, but more importantly as a zoonosis. The spores last in the soil for a very long time. Outbreaks still occur in animals in parts of Canada and the United States (often associated with bison or cattle on the range), southern Europe, and parts of Africa and Asia. Human cases continue to occur in India, Turkey, and several other countries. In the United States in 2006, a man who carried dried goat hides back from Côte d'Ivoire came down with anthrax.

It is thus no stretch to say that by 1885 Pasteur had given the world strong teeth, Happy Hours, French silk pajamas, and a vaccine for anthrax. What better legacy could a man leave the world?

So, he didn't need to turn his mind to rabies. But he did it. No one at the time knew about viruses. Scientists such as Robert Koch had barely started sorting out the relationships between the much larger bacteria and disease. But Pasteur had figured out that the brain and spinal cord were the tissues most affected in rabid animals and most able to cause disease if injected into new animals.

Taking spinal cord tissue from rabid dogs, he injected it into rabbits, which also became rabid. He then took spinal cord from the rabid rabbits and allowed it to dry. The longer it dried, the less potent was the tissue in producing rabies. Starting with the "weakest" tissue he had, he injected dogs with more and more potent disease-causing tissues until the dogs were completely resistant. But would the procedure work in people? Pasteur wasn't sure he was ready to try.

On July 6, 1885, a mother from Alsace brought her nine-year-old son to Pasteur. On July 4, the boy had been badly mauled by a rabid dog. He could hardly walk. The mother pleaded. Could Pasteur use his miracle cure? With some trepidation, he began the treatment with rabbit spinal cord that had been air-dried for fifteen days. Over the next ten days, the boy was injected (under the skin) with increasingly potent mixtures. The boy, Joseph Meister, lived. Then a shepherd came for treatment. Then hundreds more people, until, by the time I came along (much, much later!)

everyone knew about the dreaded—and very effective—Pasteur treatment. Traveling by truck, bus, train and on foot through Turkey, Iran, Afghanistan, India, and Nepal in the 1960s, I knew to stay away from strangely acting dogs. But I also knew about the fourteen painful shots in the stomach, which could save my life.

All of this seems very far away from many of us today. By the 1970s, when I went through veterinary college, a pre-exposure vaccine had been grown in duck eggs, which, I would think, was good for rabbit lovers but something of a conundrum for duck lovers. One of my classmates fainted while being vaccinated—perhaps, as an animal lover, pondering this conundrum. Now effective pre-exposure vaccines are grown in human tissue culture (another ethical dilemma). There are also effective post-exposure vaccinations, which are designed to block the virus from entering the nerves from the bite wound. These new post-exposure treatments are considerably less traumatic than the Pasteur treatment; they are called post-exposure prophylaxis (PEP), which sounds almost cheerful, with sexual overtones.

It seems that we have faced one of our darkest fears and won. We have tamed nature. For most of the world, however, this is a mirage.

For many people in North America and Europe, it is hard to imagine the situations in Guatemala, China, and India, where mad dogs and hydrophobia are not just figures of speech. Yet Guatemala, where beauty meets grief, and happy Mayans meet war and rabies, is only a holiday flight away. The beautiful old Quiche Mayan city of Quetzaltenango ("Xela"), Guatemala's second-largest city, set in a high mountain valley, does not immediately inspire terror. In October and November of 2000, rabies reared its ugly head, and four people died horrible deaths, all bitten by dogs. The Ministry of Health initiated a program to destroy stray dogs, but it ran out of strychnine and ran into protests from animal rights organizations.

The other animal carrying rabies in Latin America is the so-called "common" vampire bat, *Desmodus rotundus*. The very mention of this animal raises hackles. Vampire bats and people shared the same New World tropical ecosystems for some ten thousand years, and stories of these blood-feeding bats and the disease they carried have been part of some of the most ancient cultures in tropical America. However uneasy this relationship must have been, it was thrown way off kilter when the Europeans arrived with their cattle—blood for everyone! In the twentieth century, the main argument against vampire bats was not so much that they attacked people (which they occasionally did) but that they caused economic losses to cattle, both through causing paralytic rabies and from the effects of drinking blood.

Bats go back at least 50 million years. They are among the earliest mammals, and were here well before humanity stumbled on to the scene. They probably (mostly) came out of Laurasia, one of the two giant continents (the other being Gondwana) into which Pangaea, the original mother of all continents, split about 250 million years ago. Laurasia later fragmented into North America, Asia, and Europe.

Globally, there are about eleven hundred species of bats—20 to 25 percent of all mammals. In Latin America, there are nearly 250 species of bats; more than half of all mammals there are bats. Of the Latin American bats, about 70 percent eat insects, some of them up to five hundred insects an hour, which would be reason enough to protect them. But they may be just as important for flowering and fruiting plants, for pollination and seed dispersal. The list of plants that rely on bat pollination and spread in the wild includes everything from bananas and avocados to dates, figs, peaches, mangoes, tequila, and durian, although the last of these is, in my books, a somewhat dubious distinction.

Some bat pollinators can be considered "keystone" species: if they are eliminated, a cascade of extinctions could devastate the ecosystems in which they live. Fewer than one in a thousand, from three species, are vampires. When people first responded to vampire bat rabies, however, they did not know these things. They did not want to know. Whatever the motivation, the responses reflected the old, deep fear and rage and ranged from dynamiting, burning, poisoning, and gassing caves to leaving out poisoned bananas. Bananas for blood feeders? Clearly, the bats killed were not all vampires.

In the 1970s and 1980s, more targeted campaigns were mounted to vaccinate cattle or smear anticoagulants on them for the bats to pick up and then share with each other during grooming. In the 1990s, new patterns emerged. In some rural areas, when pigs and cattle rapidly disappeared (as many farms disappeared and livestock rearing was consolidated to respond to free trade "imperatives"), the bats looked for new sources of food and found people. In other cases, when people moved into wilderness areas—as in mining camps in the Amazon—and the bats' usual wildlife sources of food disappeared, they again looked to the delicate-skinned gourmet food—people.

Thanks largely to the works of scientific humanists like Pasteur, this dark, uncontrollable wildness at the heart of nature seems far away for many people in North America and Europe. In these places (with the exception of the Russian Federation), almost all the rabies reported is from wildlife (foxes, skunks, raccoons, mongooses, bats) and only in a few cases from domestic animals such as cats, dogs, or cattle. Human cases are rare in the United States and Canada, and for an increasing number of them, either we don't know how people got it, or the route of exposure was unusual.

Most years, no one gets rabies in Canada, the United States, or Europe. For the United States, 2004 was an exceptional year: four people died after receiving organ transplants from an infected donor. Two others were bitten by rabid dogs while outside the United States, and one person was infected from a bat but (unusually and perhaps miraculously) survived without having received vaccination ahead of time. When people do succumb, often weeks to months after exposure, it is indeed a terrible way to die. In Canada, there were only four cases of human rabies between 2000 and 2012; two were from exposures outside the country, and two were from domestic bats. In the U.S., there are one to three cases annually, most from exposure to bats, but also from raccoons, dogs, and still, very occasionally, from organ transplants. No matter the source of the virus, or route of exposure, the disease ends with hydrophobia, convulsions, salivation, asphyxia, and death.

One can understand why WHO leads a "United Against Rabies" program, the aim of which is "Zero human rabies deaths by 2030." I can even almost understand why, in Alberta in the early 1950s, the government started an all-out war that left in its wake the bodies of about 50,000 red foxes, 35,000 coyotes, 4,200 wolves, 7,500 lynx, 1,850 bears, 500 skunks, and 164 cougars. If it was not an act of science, its ferocity betrayed a passion that deserves, I think, respect.

In the late 1980s, Ontario started one of the biggest public health programs directed at rabies in the world. Many physicians and public health workers will be surprised to hear this information, because they have the mistaken notion that public health has something to do with the delivery of medical care. This supposition is only rarely true. When I began working as a veterinarian in Ontario, in the 1980s, there were fifteen hundred to three thousand cases of rabies per year in the province. This was one of the highest reported rates of any jurisdiction in the Western Hemisphere. I hasten to add that, unlike the situation in China, these were *non-human* animal cases.

Shortly after I arrived in Ontario, I visited a dairy farm to check on a cow that was "off her feed." She was a normal-looking black-and-white Holstein. She just wouldn't eat, and she was a little wobbly in the hind end. In most places in the world, one might think of some physical injury or indigestion or perhaps a toxin. Then I had to remind myself that this was not just anywhere in the world. This was Ontario in the 1980s. If an animal acts strangely, think rabies. That is indeed what she had. It was also in this rabid context that, in August 1986, a woman delivered four cute puppies to a humane shelter in a Canadian city. What she did not tell the people at the shelter was that the pups, along with their mother, had been in contact with a fox on the farm where they lived. Nor did she tell them what her veterinarian had told her five days before, that all the dogs should be destroyed or quarantined.

The day they arrived at the shelter, two of the pups were taken to a nursing home as part of a visitation program. Over the next twenty-four hours, all four pups were adopted.

Over the next week, one of the puppies had diarrhea and vomiting and then died. A second also got sick and died. Although neither puppy was afflicted with a great rush of biting madness, both, it turned out, had rabies. A total of 139 people were required to have PEP. This disproportion between the event (exposure to one animal) and the post-exposure vaccination numbers is fairly common; 665 people needed PEP after exposure to a rabid kitten in New Hampshire in 1994, and 400 people were treated after a goat at a county fair in New York came down with rabies in 1996.

No one at the nursing home or shelter actually got rabies, whether because of luck or good management or PEP we can never be sure. The mother dog remained healthy through all these events. One of the local physicians, drawing on the deep fear I alluded to earlier, pontificated that this story showed why animals should not be allowed in nursing homes. But that is a simplistic focus on disease and ignores health completely. The appropriate lesson is that all relationships carry benefits and risks and that understanding can reduce but not remove risks. Andrea Ellis of Health Canada and I developed a little booklet, *Good for Your Animals, Good for You,* which sets out rules for people who want to use animals in schools and nursing homes. The rules relevant to rabies are simple; wild animals that are friendly to people should be assumed to be sick until proved otherwise, and animals of unknown origin should never be put into contact with people until they (the animals!) have been properly vaccinated and observed for a couple of weeks.

But risk reduction for disease is ultimately not just a personal act. It is also a function of social programs. If the amount of rabies in the populations that surround us can be reduced, then the probability that people will come into contact with it will also be reduced, regardless of their behavior. If that can be done, we don't need to fear a descent into a hell of madness every time someone gets bitten.

Although most of the cases in Ontario in the 1980s were in wildlife, public health officials and physicians had to investigate 15,000 to 25,000 animal-human encounters every year, simply because the level of suspicion remained high. Because rabies was so common in animals, every animal bite could bring the deadly disease. Some farsighted biologists decided that the problem should be addressed at its source. Most of those wildlife cases, they discovered, could be traced back to the way the virus circulated in foxes. If the scientists could control fox rabies, they might be able to get a handle on the whole situation. Killing the animals, they realized, could well cause an even bigger epidemic as new, young (and sometimes rabid) foxes moved from the surrounding areas into the vacated territories and fought with each other.

The vaccination program was based on previous experience with vaccinating foxes in Europe, a series of astute field observations of fox behavior in Ontario, very clever mathematical models, and the hard work of some long-suffering wildlife biologists. They calculated that vaccinating about 70 percent of the foxes would bring the epidemic under control. The vaccine was delivered from airplanes, and by hand, in the form of edible baits. The baits are labeled "Do Not Eat," on the sometimes questionable assumption that people are more likely to read and obey such a warning than foxes. This taming of nature has been (if I may say so) wildly successful. The number of cases had dropped to fewer than one hundred per year by the new millennium.

Now, in North America, attention has focused on rabid raccoons, which have spread all through the eastern seaboard and are eyeing the land of better health care across the river at Niagara Falls, Canada, and on bats, since it is from them—or increasingly, from "unknown" exposures, which may or may not be bats or aerosolized bat excretions—that people are getting the disease. Still, globally, dog-related rabies accounts for 90 percent of all cases in people, and dogs kill lots more people than bats even when the pooches don't have rabies.

There are some effective responses to this global situation. Dog-population-control programs and vaccines, for instance, could bring rabies under control throughout the world pretty quickly. Louis Pasteur and his intellectual descendants would say, with very good reason, that we can experiment and think our way out of our fears. Pasteur urged his fellow citizens to worship in the sacred temples of scientific laboratories. He was convinced that in so doing, humanity would learn not just to conquer disease but also to turn away from fanaticism and barbarism and to find wealth, well-being, and harmony with nature.

Given Pasteur's lofty ideals, the death of one of his most celebrated patients was thus doubly tragic. In 1940, Joseph Meister, the man whose life Pasteur had saved with his vaccine, was serving as gatekeeper to the Pasteur Institute in Paris, when the German invaders—who, like their Soviet partners, pursued an ideologically driven, supposedly dispassionate science—demanded that he open Pasteur's crypt. Rather than opening the crypt to the barbarians and thus desecrating the grave of the man who had saved, and given pleasure to, so many lives, Meister committed suicide.

Pasteur was a consummate scientist of his time. He brought the world into his laboratory, and then took the results back out to the world. He did not do his work in some secretive corporate or government laboratory, trying to make money, save face, or make a name for himself. He worked in the open and thus open to challenge. The network of scientific institutes he founded, beginning with the second Pasteur Institute in Saigon (now Ho Chi Minh City) in 1891, spans the globe.

But not all of humanity works in the laboratory; many are kept out by language and money and a scientific culture of disciplinary obfuscation and exclusion. Vaccines are useless unless they are distributed and administered. As WHO reports: "Treating a rabies exposure, where the average cost of rabies post-exposure prophylaxis (PEP) is US$40 in Africa, and US$49 in Asia, can be a catastrophic financial burden on affected families whose average daily income is around US$1–2 per person."

In 2008, my wife, Kathy, and I were in Malawi, East Africa, to help lead a workshop, sponsored by Veterinarians without Borders/Vétérinaires sans Frontières–Canada, on rabies control. The day we arrived there were reports in the papers of five people who had been attacked by rabid hyenas at a local college campus. Two were killed and chewed up immediately. One died a day later in hospital. At the time the workshop started, one person was still in hospital; his face had been partly eaten away and the doctors had grafted tissue from his buttocks on to his face. Understandably the workshop participants had a heightened awareness of the importance of rabies.

Partway through the first day, participants were assigned to small groups and asked to share personal stories about rabies. The recent hyena attacks were on many people's minds. One person said her children had discovered one of the people who had been badly chewed up. In another group, a wildlife officer lowered his trousers and showed the scar where he had been attacked by one of the rabid hyenas. He and a police officer had been out hunting the animals when one of them attacked, which is unusual behavior for hyenas. The men fired their guns at the beast and it kept coming, dropped, got up and charged again, dropped after being fired upon, and got up again. It finally dropped a third time just in front of the wildlife officer. He poked it with his gun and the hyena leaped up and clamped on to his thigh in its death throes. The men were out of bullets, so they had to beat the animal with their gun butts and pry the mouth off the wildlife officer's leg. He was able to get wound treatment and rabies vaccine, which takes a couple of days to work. In Western countries, victims are usually given antibodies to rabies (immune globulin) but there is a worldwide shortage and it is very expensive, and the Malawian medical staff didn't have any available.

This is, I underline, a preventable disease. A globally successful rabies control program, or even, for all practical purposes, its near-eradication as a human disease, is possible. All that is required is something approaching a sense of human solidarity and international justice.

(14)

ABORTION IN THE SANDBOX AND OTHER PET PROBLEMS

THROUGHOUT THIS book, cats and dogs and other pets make several less-than-grand entrances, usually having to do with the quality and quantity of fecal matter they produce or with birthing fluids and materials or with their fleas. Given what is now known about the microbial world that surrounds us, are these companion animals nature's Trojan horses? Why do we bring them into our homes? Why, indeed, my farm-born friends ask. Are they bringing all manner of parasites, bacteria, and viruses into our homes? No doubt. I have friends who would say that all such animals belong in the barn or outside somewhere. Certainly a dog can do just fine sleeping in the straw, and cats can live on mice in the barn. We delude ourselves if we think they *need* the decadence of our houses. Still—even being of recent agricultural heritage—I have kept cats in the house. What is going on here?

When I have wrestled with my cats on the floor, rolling over and over, finishing with a good strong bite or an exuberant scratch from a feline, some might say that I am getting in touch with my inner child. My literary friends might draw other inferences from how our cats were named, something about genre blending: the lionesque Gabe (named after novelist Gabriel

García Márquez) and the sleek, all-black Lenny (named after poet and songwriter Leonard Cohen).

Researcher James Serpell suggests that dogs and cats can express affection to their owners, but they cannot, verbally in any case, lie to them or criticize them. It seems that pets combine the benefits of human relationship with few of the threats. Anyone who has held a door open for an indecisive cat on a cold day, however, or had a dog chew up underwear, knows that human-animal interactions can be as annoying and frustrating as many human-human ones. And I have seen enough devious dogs to know that all this pets-are-wonderful stuff needs to be taken with a grain of salt.

So why, on a tight budget, when millions of children are starving in the world and environmental causes need all the financial aid they can get, do I fritter away money and energy by keeping these little beasts in the house? Are the cats ersatz therapists, or am I perhaps only toying with a substitute for the satellite dish I don't have? Do I harbor delusions of upper-class grandeur? Are cats my version of the poor man's Cadillac?

Keeping pets is a worldwide phenomenon not restricted to modern times or Western society. In 1978, a twelve-thousand-year-old tomb was unearthed in northern Israel that contained the remains of a human being and a dog. The dead person's hand was arranged to rest on the dog's shoulder. Native Fijians of a few centuries ago made pets of fruit bats, lizards, and parrots; Samoans were fond of pigeons and eels, and Hawaiian women were known to suckle puppies. Australian Aborigines kept dingoes, the Semang Negritos of Malaysia kept pigs and monkeys, and the Navajo people of North America shared their homes with cats. The upper classes of ancient Egypt, I'm told, revered dogs, cats, and various other animals. But then their priests also revered onions, so we might be somewhat skeptical of claims to exalted status in Egypt.

Apart from the eighteenth-century Ch'ing dynasty, when imperial Pekingese or lion dogs were suckled at the breasts of human wet nurses and attended by a retinue of palace eunuchs, pet keeping—until the recent upsurge in its urban, mall-based economy—does not appear to have taken a strong hold in China. Pet culture is also relatively new in Europe. During the sixteenth, seventeenth, and eighteenth centuries, when Europeans viciously brutalized each other with religious zeal, it is not surprising that they had little empathy for other species and scoffed at New World aboriginal people who kept company with various raccoons, monkeys, peccaries, tapirs, wolves, bears, and the like.

People keep companion animals because they are seen to be useful, although the reasoning behind what is seen to be useful can be twisted. Muslim tribesmen in Sumatra still keep, and value, dogs for hunting pigs, both unclean animals. The dogs are happy because they get to eat the pigs, and the people are happy because they get to bend religious taboos without qualms. There are dogs for the deaf and for the blind and horses for the lame. Or an animal may be kept as an outlet for nurturing, care-giving behavior in a society that in the workplace assigns economics as the ultimate arbiter of what is good. In this respect, keeping a pet may be an act of rebellion against those utilitarian social values.

The therapeutic effects of pets have been much talked about and sometimes studied. The most commonly cited scientific study of the beneficial effects of pets is that reported by Erika Friedmann and her associates in 1980; people who owned pets were more likely to survive a year after discharge from a coronary care unit than patients without pets. Other studies have shown demonstrable effects of pets in lowering blood pressure, improving morale, and facilitating social interaction. Nevertheless, the most influential and repeated stories of the physical and emotional benefits of pets are just that: stories. From a scientific point of view, they provide support for the greatest medical discovery of the twentieth century: the placebo effect, the ability of humans to heal themselves. The fact that pets can facilitate this process is not something to be taken lightly and would be reason enough to use pets in therapy.

In the 1990s, our research group, led by the epidemiologist Parminder Raina, studied the relationship of pets to health and health care costs in people over sixty-five. Raina found that, for people with good human social-support networks, pets had minimal effects. For those who lacked social-support groups (an enlarging population among the aged), pets provided an important buffer in times of crisis: the less human support they had, the stronger the positive effects of the pets in improving their mental and physical health. So, pets are more than just therapy, but they are not an unmitigated good. In some groups of people—those who actually don't like animals or are poor or overworked—pets can actually decrease health and morale.

Early in this century, I was part of a research team studying animals used in hospital visitation programs. Were they bringing possible zoonotic agents into the hospital, or perhaps carrying them from patient to patient, or taking them out of the hospital into the community? Although most hospitals in North America have some sort of pet-visitation programs, there were few infection-control procedures in place associated with the disease risks these programs might pose. We found that dogs who visited human health care facilities were more likely to pick up difficult-to-treat methicillin-resistant *Staphylococcus aureus* (MRSA) and the sometimes deadly *Clostridium difficile* than dogs involved in other animal-assisted interventions. Furthermore, dogs that licked patients or accepted treats during visits were more likely to pick up MRSA and *C. difficile* than were dogs that did not lick patients or accept treats. The dogs, then, were acquiring dangerous bacteria in the hospital, and carrying them into the community. Based on these studies and several related ones, Sandi Lefebvre, the lead investigator, worked with others to develop guidelines for animal-assisted interventions in health care facilities.

As a child, I used to visit my cousins on the farm during the summer and enjoy scaring the chickens and having my oldest cousin try to squirt milk from a cow's teat into my mouth. One summer, when I went home, I found that I had raised, reddish, circular marks on my arm. The doctor called them ringworm, which is not a worm at all, but a fungal infection of the skin and hair follicles common in many animals. As a veterinarian, I used to treat it in cows by using a good scrub brush, sunlight, and fluoride-containing toothpaste, which seemed to work as well as anything more "advanced" in my toolbox.

Fungal skin infections are among the most benign things we can pick up from animals. Most of these zoonoses go from other animals to people, but a few travel from people to other animals, and some can move in both directions. The 100 million or more dogs and cats in North America "threaten" us with at least thirty diseases. Even apparently innocent birds can fluff the air with intestinal bacteria and psittacosis (parrot fever) as well as with song. Pet turtles are notorious for spreading *Salmonella*, though I must confess that the turtle of my youth, Nikita, named after the Soviet leader Khrushchev, never gave me anything worse than a desire to escape from plastic dishes and leap daringly from table tops. This he did by his daring example, which finally succeeded in his escape from my attempts to domesticate him.

The good news is that in industrialized countries, except for bites and allergies, pets are rarely identified as culprits in making us sick. Even more rarely do they give us serious illnesses, despite thousands of intimate contact hours between them and humans. The most common zoonoses show up in people as general "flu-like" sickness (transient fever, headaches and muscle aches), skin problems, stomach problems, and bites or scratches. A few, like rabies, can kill you. And some are more than passing strange.

"Could Schizophrenia Be a Viral Zoonosis Transmitted from House Cats?" is the title of a 1995 article published by two respectable scientists in the *Schizophrenia Bulletin*. The logic behind this idea was half-baked, but the hypothesis was intriguing and at least plausible, albeit barely. Although I might joke (in the privacy of my home, where after a few drinks I am known to use certain language that in public would invite a stoning) that my cats drive me crazy, it seemed far-fetched to associate them with a serious psychiatric disorder.

If that earlier suggestion was open-ended and menacingly vague, a study reported in 2005, in *The Proceedings of the Royal Society*, asked a much more specific question. Could a parasite, in particular the cat-associated tiny parasite *Toxoplasma gondii*, be the causative agent of human affective disorders? The answer was stunning. J.P. Webster and a team of British and American researchers reported that rats infected with the parasite developed a suicidal attraction to cats. Furthermore, the researchers found that two antipsychotic drugs used to treat (among other afflictions) schizophrenia, as well as drugs that treated the infection itself, were capable of altering this rat behavior.

Many parasites have a definitive host, in which they reproduce sexually, and an intermediate host, in which they may undergo some development but don't sexually reproduce. *Toxoplasma gondii* makes its definitive host home in members of the cat family. That's where, in the mucoid comfort of the intestinal walls, the parasites have sex and make babies (cysts), which then move on, into the world at large, to make their homes elsewhere, in rodents, pigs, sheep, people, or whatever other animal happens to eat them. They always return to members of the cat family to have sex.

In kittens, *Toxoplasma* causes a transient diarrhea when they are first exposed to it, usually by eating infected mice. This lasts only a couple of weeks and seems to be the only time cats can pass the disease on to others. Hence we sometimes hear the admonition not to allow cats to hunt freely, lest they pick up the infection. I am considerably more concerned about the alarming negative impact on ground-nesting birds by millions of feral cats in the countryside than I am about transient diarrhea in kittens. But there is much more to it than this.

Having flopped their way out of the kitten's bum into the wider world, the little parasite cysts may, in a period of a few days to weeks (depending on temperature and moisture), develop into a stage where they can infect any warm-blooded animals, including people. Cockroaches can carry the cysts around in their guts for several weeks after feeding on cat feces; if they got sick, one might suggest that this is a fate that is well-deserved. They seem, however, just to carry the cysts from place to place, like the workhorses of the microbial world that they are.

Once inside the non-cat species, they slip out of their cysts with a little help from digestive enzymes, change into crescent shapes, and look for white blood cells called macrophages, which are supposed to eat bad invaders. The *Toxoplasma* parasite penetrates the cell and hangs out in the back seat of the cell-car, reproducing asexually as it is spread throughout the body, eventually settling into cysts in various tissues. There they await consumption by a member of the cat family. Well, they don't just wait. In many species, infection with *Toxoplasma* parasites results in abortion, stillbirth, and neonatal death; rats and cats can get reinfected by eating the dead fetuses. If the parasites are in a rat, they change the rat behavior. The rat-cat relationships described by Webster and his colleagues seem to have been an elegant co-evolution between the parasite, its intermediate host, and its definitive host. I don't know of any evidence that human attraction to cats is the result of a brain infection with *Toxoplasma*, although the world is a wondrously strange place, and I would not discount such a phenomenon.

Different forms of the parasite were described in Java sparrows (in 1900), gundis and rabbits (1908), and a whack of other species before researchers figured out that all the parasites they were looking at under the microscope were the same *Toxoplasma*. It wasn't until the 1970s that scientists discovered the trysting, let's-have-sex-until-we-drop hideouts for *Toxoplasma*: cats. *Toxoplasma gondii* are now classified as members of a phylum called Apicomplexa, which also includes dinoflagellates, tiny non-plant/non-animal living things that thrive in the damaged coral reefs and can cause serious food poisoning, but that's another story.

Toxoplasma have inside them tiny organelles, called plastids, which resemble chloroplasts, those nano-biotech machines in plant cells that enable them to use sunlight to produce bio-energy and that are probably derived from an earlier, free-living ancestor that moved in and decided to stay. In a similar fashion, *Toxoplasma* today move into warm-blooded vertebrates of all kinds and stay there for life. The genus itself goes back some 10 million years, making them somewhat older than us. Apparently there are three genetic lineages (or clones) today, which can be traced back to two parents about ten thousand years ago. They are, one might say, more numerous than the grains of sand on the beach. What the Judaic YHWH promised to Abraham, they clearly promised to the original *Toxoplasma* ten times over. So we humans—ourselves a thriving community of microbes—are parasitized, the parasites having been parasitized. A more positive spin is to say that we have learned that we function best as living beings when we recognize our interdependence and co-operate.

In people (and other animals), the parasite goes through the gut and crosses into the bloodstream, where it heads for the delectable tissues of the muscles, eyes, and brain. In one of its forms, it has a round bottom and a cone-shaped nose (a conoid), which can rotate, tilt, extend, and retract. Once in the lymph and blood, it gets around by gliding, undulating, and rotating, using its nose to find and penetrate host cells. The parasites are a quiet lot of immigrants. Most people who are infected aren't even aware of that infection. Most of those who become clinically ill might mistake the symptoms for mononucleosis: headache, fatigue, low-grade fever, and swollen lymph nodes. Some people get eye infections (chorioretinitis), which can result in weird visual effects.

Traditionally, doctors have worried most about toxoplasmosis in pregnant women. The women themselves just get the flu-like syndrome. It's the babies who suffer. Almost half the women who are infected during pregnancy either lose the babies (neonatal death or abortion) or give birth to congenitally infected infants. Babies can be born with a whole range of diseases from eye infections to pneumonia, jaundice, fevers, rashes, and severe neurological problems requiring lifelong care.

If a woman is infected before she gets pregnant, the parasites may cause a mild illness and are then kept out of the bloodstream by the mother's antibodies. In this case, the fetuses don't get infected. In other words, the babies of women who have had infection are safe, protected by the antibodies in the bloodstream; the babies of women who have no evidence of prior infection are in danger. The parasite can get in and go straight to the fetus.

As late as the 1980s, some experts recommended that we should arrange for all children to get infected. They would get a bit of flu. They would get over it. And the girls wouldn't have to worry about picking it up during pregnancy. The other options are expensive, and few publicly funded health care programs have shown any eagerness to promote them. Some countries, such as France, screen all pregnant women and warn those who don't have antibodies to keep especially clean and report any illnesses. They can then treat new infections as they come up.

Other countries do this only on a voluntary basis, since running tests on everyone is expensive, and there is always the danger of a false positive test, which would cause the prospective mother to drop her guard.

The recommendation to infect the kids was all very well until AIDS, organ transplants, and immune-suppressing cancer treatments came along. Suddenly, in the 1990s, physicians were reporting that up to 40 percent of AIDS patients were coming down with severe encephalitis caused by *Toxoplasma*. What was happening? Were these new infections? Should people who are immuno-compromised avoid cats?

It turns out that people with AIDS, as well as pregnant women, and people undergoing cancer treatments, do not need to avoid cats, although doctors who are doing organ transplants should check to make sure they aren't passing on parasites hiding in the organs. The appearances of these serious brain infections in people with AIDS were almost entirely all the reactivation of old infections. In just about every country in the world, a third to half the population has had an infection with *Toxoplasma* by adulthood. The tiny cysts wait quietly around in the tissues, minding their own business, keeping out of the way of antibodies circulating in the bloodstream, but they are by no means dead. As soon as these community police disappear, the parasites come out and celebrate.

It is true that people can get infected by inadvertently eating kitten poop that has been sitting out for several days, although no one gets infected from eating *fresh* feces. The *Toxoplasma* oocysts need at least a couple of days to get ready for their next great adventure. It is only then, after a few days, that people can get infected. Thus, there was a lot of concern in the 1980s about contaminated playgrounds and sandboxes, which kittens were using as litter boxes. That was when I built the hinged wooden cover for the sandbox in our backyard. But the problem is more widespread than urban playgrounds and household cats. Who plays in a sandbox anymore anyway? Everybody's kid's got an iPhone. Sandboxes are for, well, cats.

Cats in the countryside not only kill mice, which seems okay, and birds, which seems not to be, but also use grain bins as litter boxes. Who can blame them? Grain is such a comfy litter. Up to half of all North Americans are thought to become infected at some point, from a variety of sources. Many of us apparently get infected when we eat undercooked meat from animals that have eaten grain used by farm cats as a litter box. Although pigs have low rates of infection, we eat a lot of pork, so that seems to be an important source of exposure.

In Canada and the United States, we don't eat a lot of sheep, but sheep are raised under natural conditions, which makes not only the sheep happy but also the mice, cats, and parasites. In a study I did of Ontario sheep farms in the late 1980s, I found that almost all of them were infected. Among the factors that determined how common the infection was in the flock were access to flowing water and the number of kitten litters born on the farm every year: more kittens means more infection. At the time of the study, I wasn't quite sure what to do with the flowing water bit. Only one outbreak of water-borne toxoplasmosis had ever been reported, in American troops who drank from a jungle stream in Panama; they came down with fever, chills, and swollen lymph glands.

Then came 1995. Dr. Andrew Burnett was an ophthalmologist in Victoria, British Columbia. In January of that year, Burnett saw a forty-nine-year-old man with floaters in one eye and complaints of hazy vision. Apart from an acute infection of the retina, he seemed healthy. Although the clinical disease is rarely diagnosed in British Columbia, Burnett suspected toxoplasmosis, which Raj Gill of the B.C. Centre for Disease Control laboratory in Vancouver confirmed. Over the next few months, Burnett and a fellow ophthalmologist, Stan Shortt, saw seven other cases with the same diagnosis. Gill, who was doing all the serum testing for *Toxoplasma* for the province, reported a sharp increase in positive results to Judy Isaac-Renton, a medical microbiologist at the lab.

What they were seeing, and what at the time they didn't know they were seeing, was the largest reported outbreak of this disease in history. Although only about a hundred people were acutely ill with a variety of the usual textbook symptoms, investigators suggested that somewhere between 2,894 and 7,718 people were infected.

For every disease, only a small proportion of those who are infected go to physicians, get a diagnosis, and appear in the official reporting system. This under-reporting varies between diseases and from country to country. Even in the most technologically advanced countries, the ratio of reported to actual can range from one in two to one in thousands. The degree of under-reporting is estimated from expensive, intensive follow-up of outbreaks, comparing how many people were reported initially with how many are uncovered through the intensive surveillance. There is a large margin of error in these estimates, as they are usually derived from several studies in different jurisdictions—hence the range in the estimates of the size of the B.C. epidemic. As soon as the size of the outbreak became apparent, physicians started a massive screening program for pregnant women and newborns.

After ruling out every other possible source of infection, investigators studied maps of the cases and realized that most of the infected people lived in a part of Victoria that got its water from the Humpback Reservoir. When they looked closer, they found that clusters of cases were preceded by heavy rainfall and turbid water from the reservoir and that people who drank more reservoir water were more likely to get the infection. But investigators could never prove definitively where the Victoria outbreak came from; they didn't even suspect the water until three months after the last case was detected. Still, that scenario best explained the data.

The water supply, they think, was contaminated either by wild cougars or by a profligacy of feral cats. *Toxoplasma* will have sex in any type of cat anywhere in the world, so it is no surprise that investigators found it in both feral domestic cats and cougars in the water reservoir serving Victoria. In a kind of Southern Hemisphere counterpoint to the Victoria outbreak, several hundred people in the Brazilian town of Santa Isabel do Ivai came down with toxoplasmosis in 2001. In that case, the most plausible story was that a cat gave birth to three kittens near one of the town's reservoir tanks; the kittens took up residence on the roof of the reservoir, and heavy summer rains washed the kitty poop into the water supply. People who drank the water, or ate ice cream made from the water, got sick.

There is no single, correct response to epidemics like this. If you are a cat owner who believes that un-neutered cats running free are natural and therefore good, you will likely think the cougar was the culprit in Victoria and might recommend fencing in the reservoir. If you are a wildlife biologist who believes that feral cats are a destructive, invasive species, you may recommend trapping and shooting free-running domestic cats. If you are a public health specialist—and considering the case in Brazil and the behavior of cats—you might be forgiven for thinking that there is unlikely to be a permanent solution but that a combination of strict domestic-cat regulation and public investment in protecting reservoirs and catchment areas might shift the odds against the disease.

In some ways, the prevention of direct disease transmission from pets to people seems simple and painless: get pets vaccinated and dewormed, don't let cats hunt in the wild, enforce laws that require owners to take their animal poop home with them, and wash your hands after handling an animal and before eating. Keep your dog on a leash when in public places.

But even what might seem at first to be a simple problem with straightforward solutions involves trade-offs. I once visited a sheep farm that had about twenty lovely gray cats slinking around, sleeping on the sheep. The cats, I was told, were there to catch mice that would eat grain intended for the sheep, and to eat placentas after the lambs were born. If the cats weren't there, the farmer would have a problem with biological waste disposal and with mice eating feed intended for sheep. It wasn't as if you could solve the problem by saying not to keep cats.

In Europe, as in Nepal and India, dogs are everywhere, inside and outside cafés and restaurants and bake shops (as are the parasites they carry). In North America, people (myself included) often keep cats because of their pseudo-semi-wild nature and their lack of dog-like cloying dependence on people. So putting a cat on a leash defeats the purpose of having a cat and may foster an unrealistic vision of our ability to tame nature, which may be translated by policymakers into ecologically catastrophic wildlife management schemes. Some people, myself included, are willing to go to great lengths to justify irrational behavior.

In North America, bites are probably the biggest threat to the public from dogs and cats. Although the fear of being bitten may draw on the deep scar left by rabies in the human psyche, the risks associated with bites are not—in North America at least—related primarily to rabies.

When I was just a little younger and still had illusions of immortality, I would sometimes go jogging in the evening to prod my body into going the distance with me. On one such evening, as I jogged through the gentle fog of a Vancouver school yard, I noticed out of the corner of my eye that an elderly gentleman was walking his dog across the shadows of the baseball diamond. What I did not see, until the dog had sunk his teeth through my trousers, was that the dog was not on a leash.

It's difficult to get good statistics on dog bites, but one study has estimated that, annually, about 4.5 million Americans are bitten by dogs, from which thirty to fifty people die. In Canada, one report suggested that forty-two people are bitten every hour, which would translate into tens of thousands every year; between 1989 and 2017, there were forty-three deaths from dog bites in Canada. Most of the more serious bites are sustained by children under fourteen. Kids haven't learned the rules of interspecies interaction very well and may become the objects of displaced aggression when they get into the middle of a dogfight or among male dogs going after a bitch in heat, or they may be perceived by the dog as a threat to other members of the dog's pack (that is, the adopted human family).

Most incidents of dogs biting people are, from the dog's point of view, understandable. The dog is simply responding to behavioral imperatives dictated by belonging to a human pack: what's bigger than you is the boss, what's littler is to be hauled into line; protect your pack; protect your home territory. If your natural behavior is frustrated, sublimate. For most dogs, these are the rules—*unless they are trained to act otherwise.* Thus, when I was investigating a rabies epidemic in Java, I found that young boys tended to be the ones most often bitten. These would be the members of the pack in a most ambiguous position, not quite dominant and perhaps a rival. Studies in some countries have found that kids are bitten in late afternoon when they are on their way home from school and may be rowdy, and the dogs are both hungry and short-tempered after an interrupted afternoon nap.

Although I will not suggest a specific school of thought on training and restraint of dogs for fear of becoming embroiled in a battle of pedagogic ideologies, it seems to me, as a North American, that having a dog respond to a leash or training it to respond invariably to commands is essential for dogs and humans to enjoy a healthy life together. It's a bit like training children in the social graces.

Some of my friends in Kathmandu would no doubt disagree. Some of their dogs act as community police, frightening thieves away at night. What good is a community police dog on a leash? The more important underlying principle is that there are clear rules for dog-human interaction that both parties need to know. Dogs do not respond well to anarchy or to having the normal rules interfered with. Children may become injured, for instance, when they get in the way of dogs that are keyed up to fight or mate, and mountain hikers in Nepal have been attacked when they approach villages unexpectedly or after dark.

Cats attack people much more rarely and usually with less serious consequences. But a good cat scratch, claws fresh from the litter box, can result in swollen lymph glands and fever ("cat scratch fever"), and a bite can inject you with an armful of cat-mouth bacteria. Although a house cat is less inclined to pay attention to disciplinary training than a dog, how a cat is brought up and socialized has very important consequences for how it relates to other living beings.

Apart from rabies, only a few zoonoses are transmitted from companion animals to people through bites, and they are sufficiently rare to be left to medical journals and textbooks. Infections transmitted by fleas, ticks, or other biting arthropods, or from environmental contamination, however, are more common. The diseases transmitted through environmental contamination include a variety of parasites.

Roundworms (*Toxocara canis* and *Toxocara cati*) are common in puppies and kittens. The dog parasites are thought to be the most common and the most important as a source of human disease. The roundworms do their sexual-reproductive thing in the puppy intestine. The puppies can get it from their mothers; when the bitch gets pregnant, roundworm larvae in her tissues mobilize and head across the placenta and out through her milk. When she cleans her pups after birth, she may reinfect herself.

For the parasite, it's a great system, evolved over many thousands of years. A female roundworm can put out more than 100,000 eggs per day. A tangled bundle of those roundworms, which cause puppies intestinal discomfort, can put out 300,000 eggs per ounce of feces. The eggs develop best at between 50 and 80 degrees Fahrenheit (10–30°c) and become infective after a couple of weeks, but they can survive for up to a year under a protective layer of snow or feces and still be viable.

In this way, an individual infection, if mishandled (by not enforcing pooper-scooper laws or not treating dogs with antiparasitic drugs), can easily become a community problem. Children, exploring their surroundings through taste, may find themselves explored by nature's little visitors. If eaten, larvae of dog and cat roundworms will wander around in the human body, obviously lost in this cavernous wilderness of unfamiliar flesh, sometimes coming out the eye, where they are called ocular larva migrans, or ending up in other organs, where they are called visceral larva migrans. Many (anywhere from 8 to 63 percent of) playgrounds in North America, Europe, and Japan, as well as many backyards, have been shown to be seeded with eggs. The contamination rates in yards are not related to whether or not one owns a dog. Dogs wander and poop where they like.

In North America, where there is some modest control over dogs, scientists tend to think of this as a disease of children in unclean environments. This is not so everywhere in the world. A 1987 study in the Midi-Pyrenees region of France uncovered several dozen adults with weakness, rash, breathing trouble, and various allergic-like reactions. The researchers suggested that the disease was a "new disease syndrome" associated with visceral larva migrans and that the people had probably picked up the disease from their hunting dogs. The larvae from a variety of roundworms, however, can cause similar problems in people. In some parts of the world, these may be pig roundworms.

Domestic dogs and cats are not the only urban animals that carry zoonotic parasites. Urban centers have provided wonderful housing and free food for raccoons in North America, and rabies, though at the top of the list of worries about these waddling suburbanites, is not the only disease on the list. Besides some afflictions that are uniquely their own, raccoons also harbor a species of roundworms called *Baylisascaris procyonis*. Wildlife disease experts suspect that 70 to 90 percent of raccoons might carry these roundworms.

Children interact with their environments by tasting them, or by crawling through them and then putting their fingers into their mouths. Hence they are at greater risk for acquiring any diseases transmitted through environmental contamination. When children eat the infective eggs with which the raccoons have fertilized the environment, the lost roundworm larvae sometimes migrate to the brain. A 2005 newspaper report from Toronto described a seven-and-a-half-year-old who suffered severe brain damage even as perplexed physicians kept him in quarantine and ran all kinds of tests, including an uninformative CAT scan. Not until blood and fluid samples were tested at a veterinary laboratory at Purdue University could they make a final diagnosis. The boy survived but is on long-term therapy and rehabilitation. In North America, there have been more than a dozen known deaths from this disease, but because doctors tend not to look for it, and because there aren't any readily available, reliable tests for it, no one knows how many people are out there with raccoon roundworm larvae wandering around in their bodies.

Several of the bugs that give dogs or cats the trots can make people quite sick as well. A 1984 study of *Campylobacter* infection (which causes bloody diarrhea) in college students in Georgia found that eating chicken and owning cats were the main risk factors. It should go without saying (but it doesn't, believe me) that an animal with diarrhea should not play with little children or visit a nursing home.

The small risk of disease and injury that a pet may pose should not scare us away from our fellow creatures, any more than influenza or AIDS should drive us away from the company of other humans. For most of us, the *health* benefits of cavorting with dogs and cats far outweigh any *disease* risks. Reckless love of animals, like reckless love of the earth or of other people, carries with it natural consequences. What these diseases are telling us is that we need to "care" for each other, in the best sense of that word. To do that well, we need to gain a better understanding of the complex ecological and social roles that companion animals play, carefully managing the risks they pose even as we value their benefits.

I recall my nine-year-old son, Matthew, and me crying over the death of a young guinea pig and then, together, carrying out a post-mortem to see why it had died. Having ascertained that the cause was pneumonia, we could bury the creature in a little grave and put this knowledge to some use in how we cared for the next one. The tensions between life and death, body and spirit, eating and sacrifice, individual and group, predator and prey underlie an ecologically based ethic.

Companion animals present us with a microcosm of the inherent contradictions between health and disease; they carry parasites and microbes that could make us sick even as they lower our blood pressure and make us feel happy.

If these animals, many of whom have accompanied humanity on our long evolutionary journey from Africa, can help us to realize that we are but one of the multiple voices of nature, and that every social act is also an ecological act with possible disease-causing consequences, then they will have done us a great service. We may, after all, be remembered as part of a bright and motley choir of many species, from snow leopards, quaggas, aurochs, white-footed field mice, cicadas, cows, mosquitoes, poodles, and pot-bellied pigs to Carolina parakeets singing in this small, pale sphere of light, our song an aurora borealis shimmering across the vast darkness of the universe.

(15)

POKER PLAYERS' PNEUMONIA

THREE OR four times a week, eight men and four women gathered for a few hours in the small room, playing poker, drinking beer, talking. From a social-health viewpoint, this group could be seen as exemplary: not only did it include members of both genders, but also some were of Caucasian and some of African heritage. They were clearly building a social support group (what the World Bank likes to call "social capital"), and there is good evidence that communities with a lot of social capital also tend to have better health.

That winter of 1987 in Halifax, Nova Scotia, the poker players didn't pay much attention to the cat until Valentine's Day, when she gave birth to three mewling kittens. That was pretty cool. One of the kittens died, but these things happen. That's why cats have more than one in a litter and more than one litter in a lifetime. In so-called r/K selection theory, "r" strategists produce a lot of offspring and hope that some of them survive, and "K" strategists produce one or two and protect them with their lives; I would put cats slightly toward the "r" side (although not so far in that direction as, say, spiders).

On March 5, the cat's owner got sick. Over the next couple of weeks the rest of the poker group came down with cough, chest pain, sore throat, nausea, vomiting, diarrhea, and, not surprisingly, fatigue. Eleven of them got better either spontaneously (perhaps because of the beneficial effects on the immune system of good social capital) or after medical treatment with antibiotics (which the World Bank might call human capital, and which we use when the social capital fails). One person, who had a heart problem to begin with, died. Joanne Langley, Thomas Marrie, and the other epidemiologists at Dalhousie University who investigated the outbreak and reported it in the *New England Journal of Medicine* dubbed it poker players' pneumonia.

The disease that attacked the poker players is more widely known in the medical community as Q fever, a less interesting and less informative moniker. This is the kind of name given to a disease or agent when investigators don't have a clue what they are dealing with; thus, aflatoxin poisoning in England was first called turkey-X disease, and the strain of hantavirus that laid low the Navajo in the Four Corners region of the American Southwest was called *sin nombre* (no-name). The name Q fever came from Australian physician Edward H. Derrick, who in 1935 investigated an outbreak of fever in abattoir workers in Brisbane, Queensland.

According to his colleague Macfarlane Burnet, who with David White later wrote the classic book *The Natural History of Infectious Disease,* Derrick rejected the name "abattoirs fever" because it might reflect poorly on the meat industry. He couldn't use "X-disease" because it was already in use. Burnet suggested "Queensland rickettsial fever," but this was rejected because it was considered derogatory to Queensland. Derrick finally settled on "Q" for "query" fever.

As part of his investigations, Derrick injected blood and urine from sick workers into guinea pigs, who, as stalwart pawns in the army against disease, also got sick. Unable to isolate any agent, Derrick took a saline emulsion of sick–guinea pig liver and sent it to Burnet in Melbourne. Burnet, later to win the 1960 Nobel Prize in Physiology or Medicine for his work in immunology, isolated the tiny organisms. They looked, he thought, a lot like rickettsiae, which are somewhere between bacteria and viruses in size and prefer the comfort of living inside cells to the rough-and-tumble life in the free world. If they were capable of thinking and writing, one might suggest they first penned the words later paraphrased by the poet Leonard Cohen, that the cell-free life might look like freedom, but it feels like death. If, as Lynn Margulis has suggested, we are, in whole or in part, communities of bacteria, Cohen's lyrics take on a new layer of meaning.

At about the time Burnet was ferreting out the tiny organisms in Australia, Herald Cox and Gordon Davis at the Rocky Mountain Laboratory in Montana were studying ticks and the organisms they carried to try to understand two other zoonoses, Rocky Mountain spotted fever and tularemia, the latter a plague-like disease also called rabbit fever and deerfly fever, caused by a bacterium, *Francisella tularensis,* and transmitted to people through ticks, deerflies, or direct contact. A rare disease (about a hundred cases in the U.S. per year), tularemia is now considered a potential bio-terrorist agent, which to me indicates an unprecedented level of paranoia. Cox claimed to have cultivated rickettsiae from the ticks, but Dr. Rolla Dyer, director of the National Institutes of Health, was skeptical. Dyer visited Cox to challenge him and came down with fever, chills, sweats, and pain behind the eyeball (retro-orbital pain). In one of those rare instances where poetic justice does occur, blood from Dyer also made the guinea pigs sick. After the usual chaotic name calling that characterizes much microbial classification, investigators seem to have settled on the names *Coxiella burnetii* for the organism and coxiellosis for the disease (although "Q fever" is still widely used).

Q fever is now reported from Uruguay to Portugal, from Russia to Australia, from Canada to China, from Switzerland to anywhere there are sheep and dust.

What was peculiar about the work of Thomas Marrie and his friends in Nova Scotia was that the disease was associated with cats. I have a paper in my files (somewhere) that reports quite definitively that cats do not get Q fever. I also have notes from veterinary school that say quite definitely that prions do not cross the species barrier, demonstrating not that such pronouncements prove science to be a bad thing but that science does progress and that scientists should be circumspect in their pronouncements. In this way, however, scientists are at a disadvantage against the pope, archbishops, grand ayatollahs, rabbinical teachers, evangelist preachers, and the presidents and prime ministers of many countries, who appear to have all the final answers.

Marrie went on to investigate the disease in the Maritimes and to review what is known globally about this zoonotic condition. He discovered that about 20 percent of community-acquired pneumonia in rural Nova Scotia was related to cat exposure, but he also investigated an outbreak related to skinning (dead, of course) wild rabbits. According to his team, almost half the snowshoe hares in the region had antibodies to *Coxiella burnetii*. One might suggest "wild-rabbit-skinning disease" as a good name, so as not to unfairly denigrate poker players or cats, but "Q fever," being shorter and more neutral, may be less open to the political vagaries of public language.

Sheep and goats are considered the most common reservoirs for *C. burnetii*, and in a study our group did in Ontario in the late 1980s, we found that pretty well all flocks had some evidence of infection. Outbreaks and "abortion storms" associated with coxiella infection continue to occur in sheep and goat flocks around the world, but most animals carry the microbe without showing signs of illness.

This cryptic infection has been known for decades, and it was with some dismay and bemusement that I read reports that sheep were being taken into the Hospital for Sick Children in Toronto in the early 1980s, to be used for research. The sheep were dropped off in the basement of the nurses' residence, then carted through a tunnel into the hospital itself and taken up to the ninth floor in an elevator also used by patients, visitors, and staff. As many as twenty sheep and a few goats were up there at any given time. If the research animals needed surgery, they were taken down to the eighth floor and sometimes walked down the hall. Nobody had reported getting sick, but an investigation found at least a dozen people who had had some variation of fever, diarrhea, coughing, muscle pain, and in one case hepatitis, that could be attributed to infection with *Coxiella*. The researchers almost seemed surprised.

Not long afterward, I phoned around to medical schools in southern Ontario to find out what kinds of courses were being offered on the epidemiology of zoonoses. Several thought such a course might be a good idea and had never heard of one that had been offered at Guelph for twenty years; diseases from animals were way off the medical radar.

Not long after the investigation at SickKids, and at about the time an outbreak in people who had contact with sick goats at the Royal Agricultural Winter Fair in Toronto was unfolding, one of the few veterinarians in the Ontario Ministry of Health was accosted by a new, perplexed minister of health with the question: what's a veterinarian doing in the Ministry of Health? What indeed.

The late 1980s were a time of generally heightened concern (sometimes verging on panic) about Q fever and sheep. The incidents at Toronto SickKids and community outbreaks stimulated a review of how sheep are screened before entering research centers in Canada, including the Ontario Veterinary College, where I worked. Veterinary students learn a lot of important things from sheep, although not how to panic and run at unexpected noises, which is a good thing for future clients. For four years, in parallel with the procedural changes for disease-screening experimental sheep, I ran a "blood-for-doughnuts" program. In return for doughnuts and coffee, students in one veterinary class donated blood every semester and filled in little questionnaires about exposure to animals. I followed the class from their first year all the way through to graduation, but only one or two of them showed any evidence of exposure, and that wasn't at the vet college. One could almost hear an administrative sigh of relief.

But sheep and goats aren't the only reservoirs, as Thomas Marrie demonstrated. One of my late colleagues in Ontario, Gerhard Lang, who himself suffered from a chronic form of Q fever, found widespread evidence of exposure to the agent in cattle as well. There is evidence that infected cattle can shed it in their milk for more than two years, but, although the cattle in Ontario had antibodies, indicating that they had been exposed, there was no evidence that they were actively infected.

Researchers and veterinarians don't have a very good idea which animals might be carrying and shedding the organism; it is very infectious at small doses (one organism is apparently enough to make a person sick), and laboratory technicians are not keen to work with it. We are therefore left with secondary evidence, such as antibodies in serum, which indicate that the animals were once exposed to the organisms but don't say much about current infection or shedding.

In the absence of abortions and stillbirths in animals, there is not much to go on, either in animals or people. Clinical signs (things that can be seen and measured, such as temperature) and symptoms (things that are felt, such as pain and headaches) aren't unique to Q fever. The disease also varies from place to place, probably depending on particular strains of the organism. In the Maritimes of Canada, pneumonia seems to be the main disease; in other places, it is hepatitis, diarrhea, or general "flu-like" syndromes, which is why it is not surprising that the hospital-infection-control people in Toronto had not detected an outbreak until after it had occurred. Chronically, *Coxiella* can stick to heart valves and cause a debilitating endocarditis.

I have left one question hanging in the air: how do people get the disease from animals? In the summer of 1981, twenty-nine people in Gwent, south Wales, came down with fever, severe headaches, and malaise. "Malaise" is a technical term for feeling bummed out, under the weather. (I sometimes ponder, as a specialist in zoonoses, what wonderful afflictions I might be suffering. Alas, I suspect my malaise has more to do with newspaper headlines than bacteria, viruses, or rickettsia.) Two of those afflicted in Gwent county developed something considerably more serious than my middle-class, hypochondriac whinging: hepatitis and endocarditis. The person with the infected valves died, and the cause of death was demonstrated by infecting and then sacrificing a guinea pig; those who say animal sacrifice is obsolete little know the costs of our well-being and scientific knowledge. Investigators followed every possible lead as to how people got infected. In the end, they decided that farm trucks had driven through the neighborhood, stirring up clouds of contaminated straw and manure.

At least the good people in the villages of Val de Bagnes in Valais Canton, Switzerland, enjoyed the pleasures of watching sheep, and not just ramshackle trucks, flock down their valley from mountain pastures to market. The autumn of 1983 was an especially dry one, and over four hundred villagers came down with Q fever. About half the people had shivering, severe headaches, severe exhaustion, and loss of appetite. Some of them had more severe chest pains, coughs, dizziness, and abdominal pains.

According to the investigators, "More than half of the patients suffering from Q fever were not seen by a doctor, being either only slightly ill or not ill at all." This is an interesting use of the word "suffering," which I shall have to keep in mind when tallying up a lifetime of such afflictions. On the plus side, the researchers made the remarkable suggestion that physicians and veterinarians talk to each other and perhaps even work together. After decades of resistance, diseases such as Q fever, avian influenza, and West Nile virus have encouraged some movement in that direction.

What almost all the outbreaks have in common is that fluids and materials from animals that have recently given birth spill into the environment, dry up, and get blown around by the wind. They are tough and can survive a long time under difficult environmental conditions. Near Aix-en-Provence, France, investigators attributed higher rates of the disease to a local wind, the mistral, which came down from the northwest and passed over a sheep-rearing area before entering the small town of Martigues; the pristine air of southern France was harboring more than romance for starry-eyed tourists. Some cases have been traced to drinking infected milk, but I suspect that the milk splashed and people breathed in the fresh, warm, down-home scent of infection. One version of the infection cycle shows *Coxiella burnetii* being transmitted from wild animals to people and domestic animals by ticks, which, given all the other bad stuff ticks carry, should not be surprising.

In 1978, four employees of an "exotic bird and reptile importing company" in New York State were set back with the usual fever, chills, and severe headaches. They had been unpacking and deticking a shipment of five hundred ball pythons (*Python regius*); the ticks were carrying *Coxiella burnetii*. Carriage by ticks is not thought to be common, but then who would have thought pythons were pets? The people who shipped the snakes from some tropical country must have shaken their heads in wonder; maybe they thought Americans eat them? The traffic in such so-called exotic pets is one of the most efficient ways for strange diseases to travel the globe.

In Poland, epidemiological patterns and experiments in mice suggested the possibility of sexual transmission, which could lead to all manner of inappropriate jokes. But I suppose a great many infections can be spread that way, depending on how one defines "sex."

Many of us studying zoonotic diseases had begun to relegate Q fever to that "mostly of academic interest category" when, between 2007 and 2010, 4,026 human cases were reported from the Netherlands. Responses in the Netherlands involved culling tens of thousands of sheep and goats from flocks that tested positive, prohibition of breeding of sheep and goats, compulsory vaccination, and public education. There was a shortage of vaccine, however, and the vaccination program could not be fully implemented. Prohibition of breeding small ruminants was a major hardship for those whose livelihoods was based on selling sheep and goat milk. In February 2010, the Ministry of Agriculture, Nature and Food Quality in the Netherlands, together with the European Food Safety Authority and the European Centre for Disease Prevention and Control, hosted a conference titled "One Health in relation to Q-fever, in humans and animals." The conference, which brought together experts from countries around the world, was remarkable in its holistic orientation, involving people from a variety of disciplines and jurisdictions, trying to pull together the best available global knowledge. Nevertheless, the absence of ecologists and environmental scientists, reflecting as it did the common practice of marginalizing natural scientists and ecologists from health management, made it a missed opportunity for advancing our understanding of the complex links between social and natural ecology—and helping us to get at some of the roots of the twenty-first-century "epidemics of epidemics."

(16)

BANG'S DISEASE AND THE WHITE PLAGUE

ANY CONSIDERATION of zoonoses would not be complete without considering two of the oldest and most troublesome diseases associated with domestic livestock: Bang's disease (brucellosis) and the white plague (tuberculosis).

Diseases caused by mycobacteria, a family that includes leprosy as well as various forms of tuberculosis, are among the most ancient of zoonoses. Spinal damage typical of some forms of tuberculosis has been found in Egyptian mummies from as far back as four thousand to five thousand years ago.

When leprosy all but disappeared from Europe after the Black Death swept through in the mid-fourteenth century, its sister, tuberculosis, appeared to rise up in a wave and displace it, peaking among the ill-housed and ill-fed industrial workers in the nineteenth century, dropping well before Robert Koch discovered the tubercle bacillus, and continuing to drop as nutrition and housing improved in the early twentieth century, partly in response to the bacteriological discoveries of people like Koch and Louis Pasteur.

Tuberculosis has a complex social and ecological history. The disease conjures for some the romantic "consumption" of nineteenth-century wraiths and doe-eyed poets. Others think of sanitariums where tubercular people retreated for fresh air in the first part of the twentieth century—now long gone or used as retreats for aspiring writers. Others recall the image of a foolhardy and compassionate Dr. Norman Bethune, kissing a woman with TB so that she could feel some human warmth before she died.

Except for veterinarians, few people think of tuberculosis as a zoonosis. In 1901, Koch stated absolutely that tuberculosis could never be transmitted from cattle to people, and that public health measures to prevent such transmission were ill-advised. He held to this position in the face of a great deal of contrary evidence. Much to the dismay of some veterinary scientists, such as Daniel Elmer Salmon (after whom *Salmonella* was named), who was the director of the U.S. Bureau of Animal Industry, Koch's arguments were used to support the sale of meat and milk from animals with TB.

In the next few decades, investigators from the Ministry of Agriculture in the United Kingdom demonstrated that 40 percent of dairy cattle in that country were tuberculous, and that 2,500 people per year, most of them children under five, were dying from bovine tuberculosis.

The pulmonary form of the disease with which many of us are familiar, in which the lungs are invaded and victims cough and spit blood, is transmitted by people spraying bacteria at each other. However, tuberculosis of the bones and intestinal tracts, which is more common in younger people and in developing countries, probably comes from ingestion of organisms through food and could well be from cattle. Scientists don't usually know which TB is coming from people and which from other animals; even when TB is diagnosed, the countries where it is most common don't have the facilities to differentiate among the various strains.

The findings from the U.K. in the early twentieth century stimulated disease eradication programs in cattle. Along with test-and-slaughter programs, in which infected animals (or infected herds) were slaughtered, the advent of pasteurization all but eliminated this route of infection in industrialized countries. However, the disease is still widespread in cattle (and other species) in many poor southern countries.

Mycobacteria are still pretty much everywhere, albeit in scattered pockets of poverty or wilderness. In North America, bison and elk are known to be infected. Besides *Mycobacterium bovis* and *M. tuberculosis, M. avium* occurs in a wide range of birds, as well as in pigs, cattle, deer—and in people with AIDS. One strain, *M. marinum,* is found in fish. In the United Kingdom and Ireland, badgers are infected with *M. bovis,* and in New Zealand, brushtail possums (*Trichosurus vulpecula*), ferrets, wild deer, and wild pigs. There is also evidence that dogs, cats, and parrots can get TB from people or cattle.

Given the widespread wildlife reservoirs, the difficulty in treating TB (six months of daily, expensive treatments with some potentially toxic drugs), the pandemic of AIDS, and the lack of political will to tackle difficult problems of economic disparity, TB is likely to be with us just about as long as we ourselves manage to stay around. The vaccine, BCG—bacillus Calmette-Guérin, named for the French scientists who developed it—is of uncertain efficacy; originally derived from cattle TB by a veterinarian and a physician, ironically it is even less effective in cattle than in people.

Drugs and vaccines provided the tools to finally get rid of it, but TB dug in and held on in areas of poverty and overcrowding, saved in part by misguided politicians and physicians, who now saw diseases as only medical problems and therefore saw no need to work for improving social conditions as a fundamental condition for good health. The collapse of the Soviet Union, and the general skepticism among international financial institutions about the value of government-sponsored health-and-welfare programs, along with the spread of AIDS in fragmented economies and marginalized populations, provided ample opportunities for the white plague to make a comeback.

The 2019 WHO annual report on tuberculosis estimated that, in 2018, 10 million people fell ill with TB, of whom about 1.5 million died, many with drug resistant forms of the bacteria. TB is the leading cause of death of people with HIV.

BRUCELLOSIS, A DISEASE that causes abortion in animals and fevers in people, has historically been difficult to distinguish from the background noise of the aches and pains of our normal mortal coil. There is archaeological evidence of brucellosis going back to the fourth millennium BC in Italy, Egypt, and the Near East, and to 400–230 BC in pre-Roman Britain. Since goats—one of the animal reservoirs for this disease—have been domesticated for about nine thousand years, this news should not be surprising. Also, since the disease was known locally in many areas before the cause had been identified, we should not be surprised that it goes by many names: Cyprus fever, undulating fever, Malta fever, Mediterranean fever, relapsing fever, rock fever, and Bang's disease.

The bacterium that causes brucellosis was isolated in 1887 by David Bruce in Malta. Bruce, also of *Trypanosoma brucei* fame, was investigating Mediterranean fever, first described by Hippocrates in 450 BC and in the nineteenth century a problem for the Royal Navy. The disease was a chronic, debilitating condition affecting Royal Navy seamen (and other local people, but, as ever, it was the warriors who "mattered").

Themistocles Zammit, a Maltese physician-archaeologist and a member of the British government commission run by Bruce, discovered the natural host for the disease. In the fashion of all good scientists, Zammit made his discovery serendipitously. He was planning to use goats as experimental animals, since they didn't seem to get sick from the organism. Raw goats' milk had, until then (and in many places, until now), sometimes been used as a sort of natural medicine or comfort food. Zammit demonstrated the organism in the milk of apparently healthy goats. Bruce called the bug *Micrococcus melitensis*. This organism, later renamed *Brucella melitensis*, is found almost entirely in goats and in people who handle aborted goat fetuses or drink raw milk or eat fresh cheese made from milk from infected goats. Aging cheese kills a lot of bacteria, as the cheese gets more acidic over time.

In 1897, Danish veterinarian Bernhard Bang found a similar organism, which eventually was christened with the somewhat less mellifluous name *Brucella abortus*. At least two others are now recognized: *B. suis*, found in pigs, hares, reindeer, musk oxen, and caribou, and *B. canis*, found in, well, canines.

The disease in people can start weeks to months after they have ingested the organisms and can include any or all or none of fever, sweats, depression, headache, backache, vomiting, diarrhea or constipation, certain kinds of arthritis—especially sacroiliitis—and heart, lung, and urinary tract problems. In animals, the organism causes abortion in females and inflammation of the joints and genital organs in males. It can settle into udders and lymph nodes and be excreted with the animal showing few signs of illness.

The tactics used to get rid of brucellosis in livestock have been very similar to those for TB. In this case, the eradication programs had the head start of a good live vaccine—good enough not only to protect cows, but also to inflict collateral damage on a lot of veterinarians who accidentally injected themselves while trying to jab the unhappy cattle. Once the level of disease had been decreased through vaccination, livestock farmers pushed through with a test-and-slaughter policy.

As for TB, pasteurization all but eliminated brucellosis in people except for abattoir workers, farmers, veterinarians, and all those hundreds of millions of people around the world who like raw goats' milk and fresh goat cheese, or who can't afford or who don't have access to pasteurization or firewood for cooking the milk. Also, as for TB, wildlife reservoirs persist.

In Canada and the United States, from at least the late 1980s on, various scientists, advocacy groups, and politicians wrestled with what to do about infections with *B. abortus* in wild wood bison in Wood Buffalo National Park in northern Alberta and bison and elk in Yellowstone National Park. Proposed solutions ranged from killing all the bison and putting in "clean" ones, to carrying out some kind of test-and-slaughter program, to building fences around areas the size of Switzerland, to doing nothing. The kill-all-and-repopulate strategy (which many cattle ranchers supported) raised a lot of hackles, given how poorly we understand even the simplest ecosystems. First Nations groups in Canada were skeptical of government intentions and suspected that the disease problem was exaggerated. By 1990, the issue, once raised as urgent, was dropped as being too difficult to resolve. In 2015, a paper in the *Journal of Wildlife Diseases* suggested that it was time to revisit the problem. In 2020, at the time of this writing, the do-nothing side seemed to be winning. It is not clear that brucellosis has any impact on the bison populations; although a female may abort in her first year, she is okay (and immune) in years after that.

In the U.S., hunters and outfitters have supported the setting up of feeding grounds for elk, but those also provide increased opportunity for diseases to spread. Bison and cattle don't mix much, and some have suggested that vaccinating the bison could get rid of the disease, as long as the elk populations were allowed to fall back to "natural" numbers (no feeding grounds).

When the issue first hit the headlines, I remember some people arguing that we absolutely had to do something, and we had to do it soon, or else... I don't remember where the "or else" led, but probably to reinfected cattle herds and loss of international trade, plus dire public health consequences. But no politician could stomach a public massacre of bison, so the "problem" was left. Both TB and brucellosis shuffle around in these populations, spread, and probably infect some people—hunters, mostly.

How important is brucellosis? It depends on what criteria you are using. Who gets sick is as important as how many. As historian William McNeill points out in *Plagues and Peoples,* epidemics among armies have changed the course of political and military history. So the British in Malta had good cause to be concerned.

Sometimes one case has amazing implications. In 2006, I visited the Sagrada Familia ("Holy Family") basilica in Barcelona, Spain, the organic, wildly celebratory work by the Catalan modernist architect Antoni Gaudí. According to some of his associates, Gaudi lived on a spartan diet of olive oil, nuts, lettuce, chard, bread with honey, and a few splashes of goats' milk. In 1911, the splashes of milk caught up with him in the form of a severe bout of brucellosis. This chronically debilitating disease was one of the influences that led him to set aside all other projects and focus exclusively on what was to become his (unfinished) masterpiece, the Sagrada Familia. Without the disease, would the world be culturally poorer?

In the U.S., there are maybe a hundred human cases a year. In Canada, there are far fewer. Brucellosis is described as being "endemic" in parts of Africa, Latin America, the Mediterranean, and Asia, but "endemic" in this case can refer to anything from less than 0.01 sick people per 100,000 to over 200 per 100,000. No doubt the effects outside North America are far greater than within.

There are programmatic costs to maintaining a situation where there is always some risk, where the disease is not eradicated, and where the bacteria live on in pockets of wildlife, but those may be the negotiated costs of living in a complex, beautifully heartbreaking planet. That's gotta be worth a few bucks.

(17)

DOG PARASITES IN THE LAND OF THE GODS

LIKE MANY westerners, I found enlightenment in Kathmandu. My gurus were a tapeworm and a Nepalese veterinarian.

I arrived in Nepal, for the third time in my life, on the thirteenth day of the month of Mangsir in 2049. I thought, in my Western naivete, that it was November 28, 1992. But our calendars, whether Julian (Western) or Bikram Sambat (Nepalese), were largely irrelevant.

Crossing the footbridge over the Bishnumati River in the chill, dark predawn, I could see a series of fires on the far bank, men and dark shapes of water buffalo appearing and disappearing as the light flickered, a scene from summer camp or Dante's hell.

As our little coterie of scientists drew closer, we could see several men grappling with one of the beasts, hammering a long, blood-covered spike into the base of the buffalo's skull. The buffalo sank into a bed of straw on the ground as one of the men moved in to slit the throat and get a good blood flow. The man stood up to stretch his back, wiping his bloody hands on his jeans. He was in a winter jacket with a turned-back wool collar and big rubber boots. The men piled straw over the body and fanned it into flames to burn off the coarse black hair, then stood back to watch. When the fires died down, they rolled the stiffening body on to its back, and one of them leaned down to slit open the belly. One man, who was barefoot, his loose slacks rolled up to the knees, stepped forward to pull the great, bulging white, blood-flecked rumen out of the body cavity. Another, in a red sleeveless shirt, sweatpants, and a woolen toque, cut open the chest cavity and pulled out the heart and lungs. The team of young men waded around in the spilled guts, stomach contents, and shit, steam rising into the chilled air around them as they quietly went about the work of preparing meals for tourists and other wealthy consumers.

I rubbed my hands together to keep the blood flowing and zipped my windbreaker closed over my heavy sweater. As morning light exhaled and condensed down from the snowy peaks and across the valley, dogs and pigs nosed in and out of the fading darkness at the periphery of this tableau, repeated every few dozen yards along the riverbank.

Scanning the scene before me, I wondered what the big black lumps were on the branches of the huge trees stretching up from the muck. Fruiting bodies of some sort, I thought. One of them stretched his wings and glided away. They were vultures, I realized with a shudder, hundreds of them, and a giant murder of crows, thousands of them. Our guide through this world of blood and shadows was Dr. D.D. Joshi, soon to be friend and colleague, a Nepalese veterinarian with a passionate dedication to improving the lives of his compatriots. He loosened the scarf around his neck, tucked back the sleeve of his white shirt, and pointed to the riverbank beside us, where piles of buffalo feet, manure, blood, and offal spilled down in an inchoate mess to the small black trickle, which was all that was left of the river. Ducks, pigs, and dogs foraged for maggots and bits of discarded bone and fat, and, down by the thin stream of water through the muck, men and women defecated and washed.

"There used to be broad, stone stairs down to the river here," Joshi said. "Somewhere underneath all that..." He waved his hand. "People would go down for their morning ablutions."

We stepped up the steeply sloped alleys away from the main slaughtering area, past several groups of buffalo awaiting the knife. Kids clambered over them; this was the youngsters' playground. We walked the alleys, noting a flow of dogs toward the slaughtering areas at the riverside and the curious pattern of how some groups of dogs would reach a cross-street and then stop, as if held back by some invisible force. We collected dog shit in little film containers and marked the slaughtering places on our tourist maps of Kathmandu. A little way up a narrow, potholed alley, the medieval, delicately carved wooden balconies leaning over us, we looked into a dank doorway. Inside, on a bed of filthy straw, several men were cutting up a carcass.

"Female buffaloes get killed inside," said Dr. Joshi. "It's illegal to kill females of breeding age. The government wants to ensure that there are enough work animals in the countryside." He didn't have to add that farmers will do what they have to do, cull the sick or lame, and sell them to butchers for a little income. But knowing me—knowing westerners—he did add: "Of course, we don't kill cows, only buffaloes, because of their different religious significance."

As we continued upward toward a small intersection, we passed a wall covered with half-dried pancakes of cow dung. Once dried, these burn with almost no smell, an important source of cooking fuel in a country with no oil reserves and an increasingly deforested countryside. The city was beginning to wake up. Inside houses everywhere, I could hear hacking and coughing and lung clearing. A great gob of phlegm was lobbed out a shop door past me and into the street. I recalled that lower respiratory disease and tuberculosis were both on the top ten killer list globally and were common in Nepal. A few seconds later, a woman in a well-worn sari stepped out, looked up and down the street, smiled happily at us, and disappeared back into the darkness of her shop.

We stopped for sweet, milky Nepalese tea at the corner, cupping the glasses in our hands to warm ourselves. Wondering if the tea were hot enough to kill the bacteria in the water used to rinse the glasses between customers, I spilled a little over the lip on the side from which I planned to drink. Soon the sky had turned from gray-pink to pale blue, a mist hanging in the early morning December air, veiling the white vistas of heaven. A rickshaw was being pushed uphill past us, and we stepped out to have a look at the pile of meat spilling over its sides.

Peter Schantz, a veterinary parasitologist from the CDC, pointed out a fist-sized, whitish, semi-opaque cyst in the lungs in the rickshaw. He rolled it bare-handed between his fingers. This was the parasite we—Canadian veterinarian Dominique Baronet (my graduate student), Schantz, Don de Savigny, from the International Development Research Centre in Ottawa (our funder, and a parasitologist himself), Dr. Joshi, and I—were here to understand and to do something about. There were a lot of problems associated with the dogs of Kathmandu, including bites from feral dogs, which might, or might not, be rabid, or perhaps just mean-tempered. But Joshi had not brought us halfway around the world to do something about rabies, as important as that was. He already knew what could be done about rabies and was already running free clinics to vaccinate dogs as often as he could get donations to do so. No, Joshi had brought us to this medieval city that was his home to help him figure out how to deal with a tapeworm of dogs.

In dogs, *Echinococcus granulosus* is a tiny, mostly harmless tapeworm. If you follow the dogs around the narrow streets of Kathmandu and carefully, painstakingly finger your way through piles of fresh feces (much to the delightful scorn of neighborhood children), or better yet (from an investigative point of view), thread your way through the yards of intestines in a freshly dead dog, you can sometimes find tiny, glistening bits of raw pasta. These are the adult tapeworms. Trying to find a living specimen is a lot like trying to find a grain of wet rice in a pile of dog feces.

Dogs pick up the infection by eating the fertile cysts of the kind we had seen in the buffalo lung; these large, fleshy balls of fluid and tapeworm heads are found in herbivorous animals of many kinds, the most important of which worldwide have been sheep. In the Arctic regions of the world, caribou, reindeer, and voles more commonly serve as intermediate hosts. The term "host" has a genteel, Emily Post resonance, as when the pathologists ask me to *appreciate* a fine lesion. But "host" refers to the larger animals on which the smaller animals (the parasites) feed.

The tapeworm is a good guest in the dog, in whom it does not appear to cause any disease or discomfort. The cysts are another matter.

Historically, as for many other multihost parasites, scientific investigators first discovered what they thought were several unrelated diseases in people and animals, each with a different story. In people, for instance, the slowly growing cyst came to be called hydatid disease. Such cysts have been described in both people and animals, going back to early naturalist-healers such as Hippocrates (third century BC), Aretaeus of Cappadocia (first century AD), and Galen (second century AD). While physicians tended to view such cysts as tumors growing spontaneously from various organs, as early as the seventeenth century some naturalists suggested similarities between the cysts in people and cysts in other animals and described what appeared to be tapeworm heads inside them.

Depending on where the cyst occurs, the clinical disease has different manifestations. Usually it settles in the liver, where, as it enlarges, it acts as a slowly growing tumor and causes abdominal discomfort. In severe cases, people look as if they are pregnant;

in an alien-species-type way, they are: pregnant with thousands of tiny tapeworms. If the parasite makes a cyst in the lungs, such as the one we fondled in the rickshaw, it can result in a dry cough, or, if it ruptures, coughing of blood. One species of the parasite (*Echinococcus multilocularis*) occurs in the circumpolar region of the globe and is common in some parts of Tibet and northwestern China. *E. multilocularis* cycles between caribou and wolves or between foxes and voles, and has been found in coyotes, foxes, and dogs across Canada. In people, it behaves more like a malignant tumor, sending little buds and daughter cysts around the body.

There are few good treatments for hydatid disease. Surgery to remove the cysts can be risky, for if the cyst ruptures, hundreds of tiny "daughters" are spread all through the body, sending the patient into shock and leading to death. Antiparasitic drugs, which have been tried in various combinations and for different periods of time, are only partly effective in reducing cyst size and require long-term treatment.

Dogs contaminate the environment with wriggling sections of tapeworm, each containing several hundred eggs. Intermediate hosts—sheep, goats, camels, water buffalo, swine—ingest the eggs, which (to simplify a more complicated story) slide down the esophagus and through the stomach and then migrate across the intestinal wall and through the bloodstream to some congenial resting place. There a cyst develops, within which daughter capsules bud off and little prototapeworms are formed.

This cyst stage of the parasite is not harmful to people. People could eat these cysts without becoming infected, though I don't know of any good recipes offhand. It's the dogs that get infected from the cysts. When the sheep or other host dies, and a dog or a coyote or a fox eats the viscera or is fed the cyst as waste from the slaughter, the baby parasites attach themselves to the intestinal wall and complete the life cycle.

In this cycle, people take the place of other livestock. This fact is a useful reminder, for those who might have forgotten, of our animal nature. Thus, we pick up the disease from dogs. In some cultures, such as among the Turkana people in Kenya, who live in an area where water is scarce, dogs may be used to clean off young children. Being licked by a dog might be better than no bath at all, but it can also transmit disease. In Lebanon, shoemakers once used a mixture of dog feces and water to prepare hides and thus were at special risk. Among certain groups of people in Kathmandu, Dominique Baronet discovered, dogs are valued members of the community for religious and social reasons. In many cultures, children snuggle up to dogs or play on floors in houses where dogs may have relieved themselves.

The unfortunate parasites that end up in human bodies find themselves stuck as strangers in a strange land, in what parasitologists call a dead-end host. People are not really a part of the natural maintenance cycle. In most cases, people don't reinfect dogs; that's the natural job of other livestock species. The reason for this is that we don't normally slaughter people; well, that's actually not true, although often we cremate them or make some attempt to hide the bodies after we have slaughtered them. More important, we aren't usually eaten by dogs. An exception may occur among the Turkana, who sometimes bury their dead in shallow graves and whose dogs are hungry.

But in Nepal, humans were a dead end. In a survey of local hospitals, Joshi found that more than 20 percent of people who had surgery for cyst removal died during or shortly after surgery. However common or uncommon the disease might be, the prognosis of those who got it was not good.

By the late nineteenth century, scientists were beginning to understand the complete life cycle of the parasite; today, dog, livestock, and human diseases are often referred to by the collective term cystic echinococcosis. This parasite has a home on every continent except (possibly) Antarctica, although the prevalence varies from place to place, and a few countries have successfully eradicated it. It occurs at highest levels in parts of Africa, central Asia, and southern South America. In 2006, a team from Switzerland calculated that the global burden of disease imposed by cystic echinococcosis was on the order of a million disability adjusted life years.

THAT DAY IN 1992, having seen the animals being slaughtered, the cysts, and the dogs, we continued up the streets behind the fires—narrow brick and potholed asphalt streets curved up between ancient brick walls and finely carved, cracked, ancient wooden balustrades. We paused at Durbar Square, an area of ancient temples, small shops, wandering tourists, open markets, and the "old" city of Kathmandu. Freak Street, which led away from the square and which I had visited as a vagabond in 1967, had slumped into a quiet backwater. The place to be now for travelers was Thamel, a twenty-minute walk from Durbar, its bustling streets crowded with restaurants, and hiking and souvenir shops, a magnet for hikers, dope dealers, carpet sellers, and seekers of enlightenment.

We circled back down the main road to the Bhimsenthan Bridge. The vendors were laying out their wares on the sidewalk: brass pots, cotton shirts, and fruit. Tailors were tuning their sewing machines. Below, the pigs foraged in offal from slaughtered buffalo and open latrines. I looked up to cloud-shrouded mountains and then along the bridge. Nearby, a girl on the sidewalk very carefully stacked lychees and peaches in pyramids of five. With nimble grace and dignity, she crossed her legs and waited.

As we walked back to Joshi's office, I noticed a small, thin dog sitting on a blanket near a door. An old Gurkha dozed in the doorway. A heap of dogs lay nearby, their bodies rising and falling in peaceful sleep. A little farther along, three small boys in rags were piled up like sacks of rice, sleeping. By this time, the sun was up and the temperature rising fast. By midday it was in the mid-seventies, and we had peeled off our sweaters and were down to T-shirts.

We were gathered around a wooden table in a small room at the National Zoonoses and Food Hygiene Research Centre. The center was Joshi's modest two-story house with a large black-and-white sign at the gate. I could almost imagine that we were a revolutionary cell of scientists ready to bring hope, enlightenment, clean water, less garbage, and an animal inspection act to the dark chaos around us.

As we settled down with our tea in Dr. Joshi's house, we knew that to understand this parasitic disease we would need to do surveys of infection in livestock, in dogs, and in people. We would need to understand something of livestock movements, since many of the animals being slaughtered in the valley came from elsewhere and hence would have been infected elsewhere. We would need to understand something about how people and their dogs related to each other. In some ways, it was a relief to have all these details to focus on. These things were doable. Good basic science. And if we understood all these things, could we not also make them better? Didn't knowing the problem take us halfway to having solved it?

Later, filled with a sense of confidence and purpose, some of us (the westerners) went out to Mike's Breakfast, an open-garden restaurant near the luxurious Yak and Yeti Hotel. Mike had come from Minnesota in the 1960s as a Peace Corps volunteer and, like many others, had stayed. His restaurant was still the best place to get great coffee, yogurt, and fruit and listen to Mozart and eavesdrop on trekkers planning an ascent of Everest.

That evening, I stopped my bicycle in the middle of the bridge to gaze up the valley. I recalled a visit earlier that week with one of the keen young lions in the Ministry of Health. I was there to inform him that we had some human rabies vaccine available, not just for our researchers, but for use by Ministry of Health people in the area. He proclaimed that Nepal had its own perfectly good vaccine and that the human diploid vaccine recommended by WHO caused cancer. He seemed annoyed that we (outside scientists) were starting a research project when the answers were already obvious. The way to get rid of both rabies and hydatid disease, he said, was through mass gassing of street dogs.

I did not have the temerity to ask how this would be accomplished. I was, after all, a visitor, and his implication was that scientific imperialism had somehow insinuated itself into the niche left open by the political imperialists. The accusation had a ring of truth, and I was left feeling angry and disheartened. How did one demonstrate global solidarity without falling into the old patterns?

But now, on the bridge, in this warm, lucid light, I again felt hopeful. The sidewalks on the bridge were crowded with hawkers selling shirts, sweaters, shoes, fruits, vegetables, live chickens, and pots. Below me, the Bishnumati River snaked its dark, sluggish way between the herds of buffalo awaiting slaughter, the pigs, the dogs, people defecating, piles of offal and bones, chaos of life and death. The narrow, medieval streets of Bhimsenthan rose up from this visceral chaos of life and death to the centers of commerce and contemplation at Durbar Square. And above the river itself, above the dark silhouettes of the near mountains, I caught a glimpse of the white, heavenly peaks. For no logical reason I could think of, it filled me with a sense of timelessness, peace, and hope. We would prove that Ministry of Health official wrong. There were better, more democratic, more humane ways of learning our way into a healthy future.

We were going to save wards 19 and 20 in Kathmandu. And if we could reclaim this small part of Shangri-La from the depths into which it had fallen, was there not hope for the whole world? Might there not be a path into a sustainable, healthy, convivial future?

IN JUNE 1995, as I flew into Kathmandu from Bangkok, my mental landscape was in turmoil. I had been reading a *Newsweek* article about the environmental catastrophe, the cesspool of garbage and muck that was the riverside in downtown Kathmandu, the area where my collaborators had lived and worked for the past three years. The comments of one of my Canadian colleagues back home rang in my ears: "Kathmandu is an ecological disaster, a write-off. We might as well let the place crash, and start over from the ashes." Was he right? I had been taught that epidemiology is an applied science, that the reason we went to all this effort was to solve real-world public health problems. Why hadn't the work of our research team made any discernible impact?

The science part of this equation involved gathering evidence. The applied part was to persuade policymakers to take up the knowledge we generated and use it.

We had certainly gathered evidence.

SINCE WE HAD begun our work in Kathmandu in 1992, Dominique and Dr. Joshi and the other researchers had doggedly pursued the demands of good, conventional, normal science. Working with our Nepalese colleagues, Dominique spent many hours documenting the behavior of dogs in Kathmandu and collecting fresh dog feces. She then put fluorescent collars on the dogs, took pictures of them, and injected them with vaccines for rabies and antiparasitic drugs. She developed a veritable mugs' gallery of the street dogs of Kathmandu. She not only probed through dog feces searching for those rice-grain-sized tapeworm segments but also, working with a new laboratory test, searched the feces for bits of tapeworm too small to be seen with even the best microscopes.

To do household surveys, we needed detailed street maps, which were easier described than found, given the warren of unnamed alleys and small courtyards that characterize old Kathmandu. It took major detective work and days of waiting in offices with the sunlight filtering down beams of dust to yellowing piles of paper for Dominique to get detailed city maps. At least here, in contrast to Indonesia, where I had worked previously, city maps were not considered military secrets. Based on her maps, and many days of walking the alleys of wards 19 and 20, Dominique was able to conduct a random survey of households in the area to see how people related to their dogs and to assess their knowledge of disease and hygiene.

The picture that emerged from our evidence was complicated. Our team was able to establish levels of infection (and reinfection) in dogs, livestock, and people. There were cysts in 5 to 8 percent of the buffalo, goats, and sheep slaughtered along the riverbank. Many of these animals came from India and Tibet, so the primary cycles of infection were probably in those countries. Only about 6 percent of the dogs in Kathmandu were infected, and reinfection rates were very low, suggesting that the disease was still primarily an import.

Infections in both dogs and people were not, we discovered, restricted to the butchering areas. Butchers did not trim out the hydatid cysts but sold them along with the meat, which was priced according to weight. Thus, it was the squeamish consumers who cut out the cysts and fed them to their dogs. Working with the veterinary clinics we had recruited throughout the city, we were able to identify dogs in "suburbia" that were thus infected. Many of the dogs were valued as pets or guard dogs.

About 60 percent of the dog owners in the areas where livestock were slaughtered were feeding raw meat, offal, bones, or cysts to their dogs. Unlike people in North America, where feeding raw meats to dogs has become a health-food fad, these people fed their dogs raw meat because they didn't have many alternatives. In whatever country one lives, feeding raw meat to dogs is a great way to spread all sorts of diseases from livestock to dogs and to the people handling the dog food.

Usually the work went on as expected. It was the surprises that gave some of the most useful information. While Dominique was working, some artifacts were stolen from a local temple. For some of the residents, a reasonable explanation of these events went like this: Canadian woman comes into our neighborhood, injecting dogs with strange drugs. The dogs, who are our community police, die. The thieves move in. Fortunately, Dominique had built up a lot of goodwill in the community, and we could continue our work despite the suspicions.

Another complication arose when we went through our test results from people. Although 35 to 40 percent of households in the slaughtering neighborhood had at least one person test positive on an initial screening test for *Echinococcus,* these tests were not confirmed by a second test done by the CDC.

Peter Schantz suggested that maybe we were looking at another parasite, the so-called pork tapeworm, called cysticercus cellulosae in its larval form. The adult tapeworm, *Taenia solium,* inhabits the guts of people, and the bladder-like cyst housing the heads of the tapeworms lives in pigs. In poor countries throughout the world, the pigs get infected when they eat human feces, sometimes the only food available to them. Where clean water is not available, people may infect themselves, with tragic results: the cysts spin through the bloodstream throughout the body and can end up in the brain; the resulting inflammation may lead to epileptic-like seizures and psychotic-like personality changes. This disease is called neurocysticercosis. Cate Dewey, a veterinarian from Guelph, spent several years studying the connections between epilepsy and this parasite in poor villages in eastern Africa. The villagers, trying to supplement their meager farm incomes, and to support AIDS orphans and to send their children to school, had been given pigs to raise and sell. Without knowledge of the parasite, adequate feed for the pigs, or easily available clean water and latrines, the disease has become a serious burden.

Parasitologist Robert Desowitz described an epidemic of burns among the Ekari people in what was once Western New Guinea, now called (by the Indonesians at least) Irian Jaya. The epidemic occurred several years after Western New Guinea was absorbed by Indonesia, and President Suharto's armies brought a gift of Balinese pigs to soften up the local population and prepare them for an influx of impoverished refugees from overcrowded Java. The pigs were infected with the cysts of *T. solium*. The Ekari, who were fond of quickly barbecued pork (a.k.a. slightly warmed dead pig), subsequently picked up the tapeworm. Lacking both refrigeration (which over time can kill the cysticerci) and ample clean water for washing their hands, they suffered an epidemic of neurocysticercosis; during the seizures, especially on cold nights, sufferers would stumble into the fires around which they slept.

It appeared that a similar situation could be unfolding in Nepal, given the widespread presence of scavenging pigs. The tests for hydatid disease were probably cross-reacting with proteins from this other tapeworm. If so, treating dogs would do no good.

We had hoped that the epidemiological studies in Nepal would, through the sheer weight of evidence, persuade people to change their ways. The solutions were, on the face of it, obvious: have health inspectors at the slaughtering areas to cut out cysts and dispose of them properly; close in the slaughtering areas, and educate the public so that the dogs didn't get access to raw, cyst-infested meat; educate the public about hand washing after handling their dogs and about cleaning up dog feces to decrease fecal contamination of streets and houses; control the dog populations. Dominique prepared a video that was shown on TV. Dr. Joshi ran public education clinics and talked to members of the government about legislation. Yet after several years of hard work and good science, nothing seemed to have changed. Why not?

VIEWING THE HEAPS of animal parts and feces, the dogs and pigs and people foraging and defecating and washing up along the riverbank, I reconsidered previous attempts to "fix" problems of animal slaughtering, garbage, and disease transmission in Kathmandu. In the 1970s and 1980s, the Danish foreign aid agency built a large, modern abattoir (just like in Denmark!), the Germans built up the garbage disposal system (just like in Germany!), many countries helped train medical workers, and public health officials sneaked around at night dropping strychnine-laced pieces of meat for the street dogs (just as in Walt Disney's nightmares!) as a way to control rabies and hydatid disease.

In 1995, the Danish-built abattoir stood almost empty, the streets were again filling with garbage, well-trained medical workers were going back to their mountain villages to eke out a living from the barren soil, and the street dogs were doing just fine, fed and celebrated by members of the community.

These programs were based on good science and apparently sound logic. What was missing was any deep sense of culture: why did people behave in certain ways and not in others?

Scientific information is assumed to be objective and globally true, but that assumption is based on laboratory studies of basic phenomena. Can there be a science of place, one that explores phenomena that are locally true?

The earliest, most successful control program for echinococcosis was started in Iceland, where sheep rearing and sheepdogs go back many thousands of years and have long been an important part of cultural life. In 1869, Harald Krabbé, distressed at an overpopulation of dogs and the widespread distribution of "liver disease" associated with them, wrote a pamphlet in Icelandic explaining the life cycle of the disease. He then freely handed it out to every householder. Given the long, dark Icelandic winters, and the fact that this pamphlet was only the third document written in Icelandic (after the Bible and the Icelandic sagas), one can imagine that the idea of hydatid disease control became deeply ingrained in the Icelandic psyche. Even with that national breadth of understanding and a commitment to eradicate the disease (which included some drastic dog-control programs, such as legally forbidding dog ownership in the capital, Reykjavik), it took almost a hundred years for Iceland to get rid of the disease.

The success of the Icelanders emboldened many sheep-rearing countries to embark on similar control programs. The success of these programs has varied; even the most successful of them have depended on a well-educated population, regulated slaughtering facilities with veterinary health inspectors, and an attitude toward dogs that allowed for eradication of free-roaming dogs. In some cases, that success has depended on the degree to which Nepalese or Kenyans or Peruvians could be persuaded to behave like western Europeans or North Americans.

New Zealand's program was instructive; it began as a combination of education, dog registration, and provision of treatment drugs to dog owners. Education was directed at children in schools and at sheepdog owners, who were brought together to have their dogs treated with a purgative so that the feces could be examined for parasites. The program was much more successful among settlers of European descent than among the native Maori, for whom dogs played a different social role.

All of the successful hydatid disease control programs took decades to make progress and required a lot of public commitment and money. I wondered if it were even possible in Nepal.

One thing was clear: scientific work that tried to address issues that were policy-relevant, science for the public good, had to be embedded in a sense of social and ecological history. Budi Tamang, one of the Nepalese researchers on our team, recounted a story about a German development organization that went to Kirtipur, a city near Kathmandu, and installed a row of new toilets. The decision about where to place them and how to design them was most likely based on a proper scholarly assessment of public health needs in the community. For some reason, however, the people there wouldn't use the toilets. When asked, they told their benefactors, "Don't worry, maybe tomorrow we will use them." But they never did.

The Germans sent an anthropologist, who moved into the community, learned the Newari language, and joined the people's daily lives. This is what he found: During the unification of Nepal, the people of Kirtipur twice repelled the invading "unification" king. The third time, he approached from the west, although they expected him from the east. There followed a massacre. Since that time, the people of Kirtipur only defecate with their bums facing west, as an act of spite. The German toilets, as you might have guessed, were oriented so that the bums faced east. Once this mistake was rectified, the toilets were used.

Even such a simple thing as having a national act to regulate animal slaughtering—which Dr. Joshi had been working for years to get enacted—was stalled by political events. When our research team had started its work, Nepal was moving from a dictatorship to democracy. Many of the people we worked with were active in the pro-democracy movement. The political scene was exciting but chaotic. Every time the proposed act was near debate or about to be voted on, the government would collapse, new coalitions would form, and often new elections would be called. The chaotic emergence of democracy in Nepal (itself a good thing) was hindering another good thing, public health.

From a system of the king's rules, Nepal seemed to have moved rapidly to a system of no rules. Now when I bicycled around the crowded valley, I saw that the once-green rice fields of the valley were strewn with new, hastily erected, already-crumbling buildings. Everything was now for sale: intricately woven Tibetan carpets (made by Nepalese labor, polluting Nepalese water, using New Zealand wool and Indian dyes), gold and silver jewelry, turquoise and opal, Tibetan Buddhist artwork, fat laughing Buddhas and elephants from India, fierce Tibetan mastiffs, cuddly Lhasa apsos, and cheap imitations of Tibetan carpets. For the "alternative" tourists, there was the wilderness for sale: mountain climbing, white-water rafting, mountain vistas, cows, ducks, chickens, and dogs in the streets. There were also gambling casinos. In a sense, the romantic idea of a good-hearted and happy people of Nepal who worshipped strange, many-armed gods was also for sale as part of this exotic landscape.

This economic activity was bringing in a lot of money. In the best of all worlds, it would be translated into drinkable water, fresh air, and adequate food for the citizens of Nepal. But in Nepal, it was translated into impossible consumer demands and piles of garbage. How could the chaos be used as an opportunity to create something new? Many politicians, as well as the slaughterhouse owners and butchers, were responding to the demand for meat from tourists and the newly wealthy in this free-for-all economy. Could they be persuaded to reinvest in the health of their communities? How could we engage them in the work so that they would see its importance? My view of the relationship between science, policy, and public practice was shifting.

IN THE 1990S, several international networks of scholars were struggling with how to understand the complexity of the world we live in and to manage it, or adapt to it, in such ways that the human species could have a long and convivial sojourn on this planet. From these networks emerged organizations that, under banners with names such as EcoHealth (short for ecosystem approaches to health) and One Health, looked for ways to integrate the health of people, other animals, and ecosystems. We talked about, measured, and promoted the health of individuals, families, and even cities. This being the case, could we not talk about the health of whole social-ecological systems?

I thought about the web of connections, as well as the facts, that our research had uncovered. They explained a lot about why some previous attempts to improve the well-being of the urban citizens in Kathmandu had been unsuccessful.

Family and cultural traditions of butchering perhaps once sustainable in sparsely populated rural areas were brought into the very different, more crowded urban setting of Kathmandu. Butchers didn't want to give up family traditions and become wage laborers in an animal-killing factory.

The economic, cultural, and family bases of human-dog relations, butchering practices, and the many dependent occupations of small-scale meat transporters and butcher shops throughout the city could not be altered simply by decreeing that it should be so or by making public health pronouncements. The street dogs were not only threats to public health but were also community police. Butchering and food hygiene practices depended not only on knowledge but also on the availability of clean water and affordable fuel for cooking, thus competing directly with economically powerful activities such as the carpet industry, which used—and wasted—huge amounts of water.

Even if the dogs could all be treated with an effective drug, it was not clear what the environmental consequences of the drug-in-feces would be. In the end, the communities involved would still be left with serious public health, economic, and environmental problems, many of which appeared to be considerably more pressing than this particular parasite. Of all the places these communities could spend what little spare cash they might have, why would they want to spend it on an anti-parasitic drug or control program for dogs? Whatever would be done to respond to this parasite would have to be embedded in a much more comprehensive program of social transformation. If ecosystem-based approaches offered a viable way of thinking about, and working with, complex public health problems, Kathmandu seemed to be an ideal case study.

At dusk, I sat nursing a beer on the roof terrace at the wonderful old Vajra Hotel, with its wood-carved walls, gardens, and traditional statuary, on the hillside up toward Swayambhunath, the Monkey Temple. I looked out over the darkening valley, the scattering of lights in this still-medieval city, the satellite dishes along the rooftops. The scent of dung and of wood fires and the clashing sounds of popular Indian and European music drifted up past me and into the cool, almost clean mountain air. If these new ideas about the health of social and ecological systems, combining community participation with systemic ecological understanding, could work here, they could work anywhere.

POSTSCRIPT

People often ask me what happened to hydatid disease in Kathmandu. I don't know, in part because it wasn't a priority for the citizens of wards 19 and 20. I do know, however, that the conditions that made it a problem have changed for the better.

In November 2001, we had the final workshop of a community-based ecosystem health project. Our research team this time included butchers, street sweepers, veterinarians, politicians, shop owners, anthropologists, and community activists.

Later, I took a picture of Dinesh Kadji, dynamic young chairman of the butchers' association, standing outside in the clean, sunlit courtyard beside the river. I told him that I thought there used to be an old water pump here. I had a picture of a woman taking water, with her children squatting around her, and, in the background, animals being slaughtered. It's right over there, he said, pointing to a clean, well-kept building across the yard, and he laughed.

The Dante-esque riverbanks I had first encountered in 1992 were transformed. Public parks with shrubs and trees and benches to sit on had been created next to the Bhimsenthan Bridge. A man and a woman took care of new public toilets, as well as the gardens. The fees they charged were shared with the municipality. Although people were prohibited from defecating at the riverbank, I could see a fellow coming up out of the riverside grass, pulling up his pants. Still, it was much cleaner than it had been a decade previously. There were some piles of garbage, but many fewer than before. There were fewer dogs, as well. The trees were empty of vultures.

Next to the public parks, farther along the riverbank, were walls of tall grasses, and small private flower and vegetable gardens. Beyond that, rows of composting cow manure and stomach contents were aerated by plastic pipes, and flocks of ducks wandered among them, feasting on the larvae. To my right, away from the river, were cardboard and plastic recycling yards. A line of water buffalo were walking slowly along the street, followed by a lone herder. The herder guided them up a side street to a corrugated metal door. I slipped through the door after the animals. Around a brick courtyard were covered sheds. This was one of the twenty or so new, mid-scale slaughtering places scattered throughout wards 19 and 20. At a freshwater tap, women and children were pumping water into their brass pots and plastic pails.

Back outside, along the side streets, I came across the perpetually running stone taps. These taps are set in washing places below ground level for general community use and are maintained by traditional religious caretakers. Once filled with garbage, some of these washing places were now cleaned out and painted and were being used.

Since 2001, Nepal has been into and out of turmoil several times. In June 2001, Crown Prince Dipendra killed most of his family in a rage during a royal dinner, allegedly because his mother had disapproved of his choice for marriage. One uncle, Gyanendra Bir Bikram Shah, who was not at the meal, stepped up to the throne of this constitutional monarchy. Until he came to the throne, mainstream politicians had negotiated a delicate relationship with the Maoist party of Nepal and were bringing this rural insurgency group, fighting for economic development in the countryside, into the political process. When the new king came to power, this fragile peace fell apart.

In 2003 and 2004, as the Maoist war heated up, I became increasingly concerned that larger events could overwhelm a decade of good work and community learning. On the one hand, there had been a bomb near the Bhimsenthan Bridge, where the young girl had once waited with her peaches and lychees, and the king had dissolved Parliament and received military aid from the U.S. to fight the "international terrorists," thus enabling him to back away from negotiated settlements. On the other hand, I received optimistic e-mails from one of our researchers, who was now heading a group in Nepal that builds on people's skills and strengths to achieve their hopes and dreams.

At one of our research workshops, the participants had been asked what stories they wanted their children to tell about them. What I hear from Dr. Joshi is that the communities are continuing their work, fighting for their democratic rights, creating stories of resilience in the midst of the battles among the tiresome capitalist, royalist, and Maoist ideologues. They have been monitoring water quality. Joshi himself is still running dog-vaccination clinics. The social-action Nepalese partners I worked with have mounted programs to improve the nutrition and health of the people in wards 19 and 20.

The people there have discovered the story they want their grandchildren to tell, and it is one of local democracy and commitment, of clean water and a thirst for knowledge and an impatience with systemic poverty and repression. I often think back to 1992 and remember the small girl on the bridge, waiting patiently with her lychees and peaches, and I hope she has a happy role in that story.

(18)

STORIES FOR THE INTERPANDEMIC

HUNDREDS OF diseases, most of which are not covered in this book, can be shared between other animals and people. I have selected only those which illustrate clearly what I think are the most important issues we are facing in this globally interconnected, multiple-pandemic-plagued world.

At first glance, understanding how infections jump from other animals to people is not rocket science. They can be transferred through insects, environmental contamination, and direct contact, which includes situations where we rub noses with them as well as when they bite us, as in rabies, or we bite them, as in food. Knowing how they are transmitted means that preventive measures are obvious, at least for individuals. None of these measures are foolproof, but all of them are reasonably effective. Washing your hands with soap after handling possibly infected packages, animals, or meat will take care of many viruses, bacteria, and parasites; so-called antibacterial soaps and lotions and sprays are unnecessary and may only encourage the evolution of more aggressive microbial populations. Wearing a long-sleeved shirt outdoors and using insect repellents will help keep the mosquitoes and ticks in check. Cooking your food will kill bacteria like

Salmonella on the surface of—and parasites like *Toxoplasma* inside—your meat. Cleaning up dog poop and disposing of it by composting, or building mini-biodigesters in dog parks, will help keep a variety of parasites and bacteria out of general circulation.

Communicable diseases, by definition, don't come as single cases, and solutions for individuals, while important, need to be embedded in bigger initiatives. Even some preventive measures at the community level are straightforward. However, as we have discovered in the midst of the COVID-19 pandemic, they do require us to elect vertebrates in political office, and for all of us to have a sense of global equity to share the necessary resources to implement them. Social distancing with solidarity is how some people are describing this balance between individual safety and collective health. Animal and human vaccination programs for rabies, anthrax, brucellosis, Ebola virus disease, and various influenzas are all attainable using our current knowledge and technology.

How to compost manure and dead animals, instead of burning or simply burying them, and some knowledge of how energy can be generated through biodigesters, which I discuss in my book *The Origin of Feces*, should be in the training manual for every person dealing with infectious diseases. Composting not only kills most bacteria and viruses but also generates useful fertilizer. Even developing a cheap, injectable reproductive control drug for male dogs should be relatively easy if any pharmaceutical company or funding agency could be persuaded to make it a priority. Cattle can be screened for various diseases, such as sleeping sickness, before being allowed into new, disease-free areas.

GLOBAL PATTERNS OF zoonoses reflect complex social and ecological changes that go well beyond individuals and communities. It's messy, but really, we know what needs doing, and have for a long time. If we look over the natural history of zoonoses and their emergence or re-emergence in the early twenty-first century, the general causes are a mixture of things. Often humans have created new urban or agricultural ecosystems in the ruins of older, non-human-dominated ecosystems. Ebola, Marburg, Chagas, SARS-COV, and SARS-COV-2 emerged in part because people invaded new territories where other animals and their microbes have lived in some rough kind of harmony for millennia.

Animals such as rats, raccoons, and coyotes have adapted to and changed human settlements, bringing with them viruses such as rabies and hantavirus, bacteria such as leptospires, and a variety of parasites. Economies of scale and monocultures in agriculture have created ideal conditions for the generation of epidemics of avian influenza, salmonellosis, and SARS-COV-2. Fast global travel and unfettered free trade have fostered the spread of epidemics. Loss of biodiversity, social inequity, marginalization of poor people, and the rapid sprawl of slums with bad housing, inadequate water, and standing sewage have created ecosystems that change the patterns of old infectious diseases and create opportunities for new ones. Climate change, much of it human induced, is contributing to the destabilization of ecosystems and the dispersal of animals and microbes into new areas. Talk of (re)creating natural harmony or ecological stability is an illusion.

In 1848, Rudolf Virchow, a renowned medical pathologist, was sent by the Prussian government to investigate the causes of a typhus epidemic in Upper Silesia. After intensive investigation, he submitted a report that recommended a program that included "full employment, higher wages, the establishment of agricultural co-operatives, universal education and the disestablishment of the Catholic church." Okay, that last bit was pushing his luck, but the rest made sense.

In 1992, the Institute of Medicine in the United States published a report on the resurgence and emergence of infectious diseases. They identified the following causes: human demographics and behavior; technology and industry; economic development and land use; international travel and commerce; microbial adaptation and change; and breakdown of public health measures. The report suggested better surveillance, vaccine and drug development, vector control (primarily through better pesticides), and human behavioral changes—for instance with regard to sexual relations and antibiotic use—as being appropriate responses. There was no mention of regulating land use or working for more equitable economic development, health insurance, or paid sick leave.

In 2008, more than 150 years after Virchow's report, WHO published a review of evidence related to the social determinants of health. The authors, no flaming revolutionaries, declared that "social injustice is killing people on a grand scale" and recommended that governments work to "improve daily living conditions, including the circumstances in which people are born, grow, live, work and age" and to "tackle the inequitable distribution of power, money and resources—the structural drivers of those conditions—globally, nationally and locally."

A few years later (in 2012), a review of progress since the 1992 report noted that new diseases, such as SARS and H5N1, had emerged since that first report, but that the most important advances in twenty years were "genomics-associated advances in microbial detection and treatment, improved disease surveillance, and greater awareness of EIDS [emerging infectious diseases] and the complicated variables that underlie emergence." So, none of the serious causes were addressed, and few epidemiologists—at least in the meetings I attended—were surprised to see COVID-19 galloping over the horizon.

If we have known for many years what needs to be done to prevent pandemics, why haven't we acted? What these reports do not explicitly acknowledge, and what is not explicit in any list of "things to do to prevent a pandemic," is that in any such complex situation, we are faced with trade-offs. There is not a single, over-riding, science-based "truth," and many times, solutions to some problems create new ones. These "wicked problems" are everywhere around us, and central to the issues we are facing with regard to preventing future pandemics.

I recall how pleased I was when the vultures disappeared from the riverbanks in Kathmandu; the missing scavengers were at first a sign of success. Later I learned that vultures were dying all over the Indian subcontinent in huge numbers; in some places up to 97 percent of certain kinds of vultures had died. At first, investigators had thought it was some kind of viral epidemic. Then they discovered that a pain-killing drug, diclofenac, was being used in large doses throughout India to keep older cattle in comfort as they died. This in itself would seem to be a good and humane act. However, the vultures that fed on the carcasses of these cattle developed kidney failure and visceral gout (a buildup of uric acid); people recognized them by the way their heads drooped. The birds died soon after. The consequences of losing this major scavenger could be devastating, as semi-feral dogs congregate around rotting carcasses, spreading rabies and other diseases. Parsis, who place their dead in Towers of Silence in Mumbai for the vultures to clean up, are wondering what will happen to their religious practices. I pondered whether the lack of vultures I had seen along the river in Kathmandu was a sign of success in controlling butchering activities or evidence of other feedback loops, the tragic consequences of the desire to keep cattle out of pain.

Pulcher threw his sacred chickens overboard because he didn't like the messages they were delivering. In some ways, the killing of millions of chickens in the face of H5N1 avian influenza virus is a similar act of defiance. A scientist—who is, if nothing else, a reader of omens—might ask why the chickens in Southeast

Asia were becoming infected. One answer is that wild birds and livestock traders brought the virus to them. How did this happen? The draining of swamps and restructuring of landscapes for human use throughout the world has narrowed the options for wild birds to nest and to rest. Hence the birds are being forced into closer proximity to each other (more stress, more shedding of microbes) and to other species (more sharing of microbes and faster microbial evolution). They are also being channeled into smaller and fewer flight pathways, many of them close to expanding human settlements and agricultural enterprises.

Traders who are trucking and shipping poultry from place to place are responding to economic policies advocated by those in power. Those policies are put into place because urban dwellers want to eat more meat, and they want to pay less for it. Economies of scale and global trade in agriculture bring down prices at the grocery store, making many foods affordable not just for wealthy city dwellers, but also for low-income people. Economies of scale and global trade also create ideal conditions for the emergence and spread of many new diseases.

When viewed at this larger scale, the "whys" and "becauses" of any particular outbreak can go out in many different directions, and across geographic and political scales, with different winners and losers and feedback loops that are weakened (as between local farming practices and local environments) or created (as between local farming practices and global markets).

An outbreak on a particular chicken farm might be related to how the birds are housed, how crowded together they are, what other species live on the farms, what kinds of "bio-security" are available, and the structure of markets. These factors, in turn, have to do with urban demands for certain kinds of food, with economics and culture, and with who has power (both electrical, which is required to manage intensive poultry barns, and political, which is required to manage economic markets). The spread of an epidemic has to do with the technical capacity for diagnosis and response, with the value of fighting cocks and show birds, with education of poor people, and women in particular, with communication among those working with the health of people and other animals, and with drugs and hospitals available at the points where the epidemics start, which are often poor and marginalized.

What I have said about avian influenza can be said as well about just about every other emerging infectious disease. Abandonment of small farms may be seen as good (by some ecologists, for instance, who celebrate expanded habitats for wildlife) or bad (by epidemiologists concerned about the spread of Lyme disease associated with increased deer populations on those abandoned farms). Hiking through the woods is both good for one's personal health (exercise, fresh air) and possibly risky (Lyme disease, West Nile). The dogs in Kathmandu were both community police and carriers of disease. At an international conference in 2006, a scientist from Turkey described how the killing of chickens to stop the spread of avian influenza resulted in outbreaks of tick-borne diseases in Turkish villages. The chickens had served as an important way of controlling ticks.

The rapidity and scale of changes associated with human activity now far exceed what we have seen before in our brief sojourn on this planet. We are sliding over the cusp of rapid climate change and environmental change. On the one hand, the openness and speed of sharing information during the COVID-19 pandemic has given the world more opportunities than ever to respond in adaptive ways.

On the other hand, the speed and scope of the pandemic, as well as the waves of floods, storms, droughts, and fires over the past decade, should give us pause. Can we get ahead by simply working longer hours and running faster? Will a quantum computer or a 5G network save us?

Can we build theoretical and mathematical models based on our data that will help us predict, or better yet stop, the next pandemic? In the midst of the COVID-19 pandemic, the models were everywhere in evidence, guiding many national and international policies. What became quickly apparent was that there was not a single scientific model and hence not one global policy. There were many models and many policies.

As long as we don't confuse the models with the world around us, we can learn a great deal from them. Some complex systems theorists, trying to grapple with this, have described the world we live in as a series of nested hierarchies. As an individual, I have physical and social boundaries and an internal set of rules by which I function. I am also a member of a family with rules and boundaries, which is a member of various communities, each with its own rules and boundaries. By virtue of the fact that I eat, drink, perspire, breathe, urinate, and defecate, I can also be described according to my membership in several nested ecological systems.

The twentieth-century philosopher Arthur Koestler referred to each level in this hierarchy as a "holon," being both a whole and a part. The environmental scientist Henry Regier calls these nested hierarchies "holonocracies"; with its resonance of "democracy" and "autocracy," this term emphasizes mutual power relationships and responsibilities. The infinity-sign (or lazy-eight) idea of panarchy used by the Resilience Alliance, and which I introduced earlier, emphasizes the interplay between change and persistence across levels and over time.

We might ask, in relation to the Resilience Alliance model, whether we are entering a creative destruction phase. If so, then we are not only facing perilous times ahead, but we also have an unprecedented opportunity for those who desire democratic, just, and ecologically sustainable global change and renewal. The danger lies in how the power relationships play out between the destruction of the old and the creation of the new. The collapse of the USSR should be a danger flag: in the turbulent passage from old to new, oligarchs and autocrats took control of resources and power, and democratic renewal was thwarted. In the midst of the COVID-19 pandemic, one can already see autocrats (both in government and in private businesses) using a model of power relationships, expertise, and administrative control appropriate for emergency rooms and disasters to seize political and economic control. For those who wish to see a resilient, just, sustainable future, the destruction that occurs in the midst of a pandemic opens a small window in which to assert ourselves, to organize, and to act creatively.

One of the big challenges for us non-autocrats is that the questions we are asking, and the future we would like to see, require us to accommodate and evaluate many perspectives. As a veterinarian, I think of this as a clinical judgment, bringing together information from a laboratory, clinical observations, epidemiologic patterns, and patient history. For those interested in health and environmental policies this is even more challenging. Even if we accept that there might be different legitimate ways of seeing the world, the "clinical decision" is a collective one. Philosophers Silvio Funtowicz and Jerry Ravetz, building on their work with uncertainty in environmental risk assessments and the social problems that complicate our understanding of scientific knowledge, have called this publicly engaged science "post-normal." Going beyond "postmodern," in which many different viewpoints merely stand side by side, they propose that many different kinds of evidence and perspectives need to be called upon to understand the world, define problems, and resolve issues in such a way as to sustain a convivial human existence on the earth. Post-normal science, they have argued, is most appropriate in situations where "facts [are] uncertain, values in dispute, stakes high and decisions urgent." At its core, post-normal science is a democratization of science, and I have yet to find a better way to think about the sorts of challenges we face with ZOONOSES, EIDS, and pandemics.

The main character in William Boyd's novel *Armadillo* is searching for a word to signify the opposite of serendipity. "Serendipity" is derived from Serendip, an earlier name for what is now Sri Lanka. Heroes in an old Persian folktale had the faculty of making happy and unexpected discoveries by accident. Boyd suggests that a term to mean the opposite of "serendipity" be based on a fictional country that is cold and barren, called Zembla. Thus, "zemblanity" would be "the faculty of making unhappy, unlucky, and expected discoveries by design." This faculty is certainly what characterizes the disciplines to which I have devoted my professional career—epidemiology, environmental, and biomedical sciences. We think some things, such as feces in water or diesel fumes in air or smoking cigarettes, might cause people and animals to get sick, so we set up studies to prove that this is so. To the surprise of no one except those who are making money by dumping this stuff into our air and water, we find they do cause disease. Much global monitoring of disease and environmental status is also characterized by zemblanity. But the problems of the twenty-first century are unpredictable, and the future, thankfully, is uncertain. To co-create a more convivial world, we need a new serendipitous science, a post-normal science, a science that is not afraid to look at everything, to accommodate uncertainty, mystery, struggles for justice, and to admit storytelling into its campfire circle of models and laboratory results. It is a delicate balancing act.

After decades of studying zoonoses, epidemics, and pandemics, and working with communities in all parts of the world to help them put that scientific knowledge to use in ways that might improve their daily lives, let me offer my story about how I have come to think about science. In the seventeenth century, during a rare interpandemic phase of the bubonic plague, and in the midst of multiple wars, René Descartes argued that, by dividing the world into smaller and smaller pieces for observation, we could "render ourselves the lords and possessors of nature," and, in so doing, improve our health. Descartes argued that people should step away from the books of the old masters, and go out to observe the world in real time. Fair enough, although most students since then have learned their science from books. Still, since the seventeenth century, Cartesian science at its best has enabled us to learn a great deal about the things of which the universe is made. In terms of COVID-19 and other pandemics, for instance, Cartesian science has enabled us to identify bacterial and viral structures, develop vaccines, and put into place protocols to reduce the spread of disease. And yet, the notion of becoming the "lords and possessors of nature" eludes us. Why might this be? Zoonoses and pandemics, and, in particular COVID-19, have brought with them questions of how we understand, and respond to, the forces of nature.

The world is not just the stuff around us. In a pandemic, it's not just the viruses, other animals, and people that matter. What holds our world together are the relationships among them. These relationships are expressed—for lack of a better word—through conversations. We know about a few of these conversations, those expressed in the sounds animals make and biochemical transmissions among insects and between insects and plants. Others, like gravity and atomic forces, we know not because we see or hear them but only by their effects. Since many of these tangled and shifting conversations around us cannot be brought into a controlled laboratory, our understanding of them is at a very rudimentary stage.

What we struggle with, as people, and, more specifically, as scientifically trained scholars, is finding a language that can not only encompass complex uncertainty but enable us to engage in it more fully. Until now, we have learned to lecture the living things with which we share the planet using the very blunt language of bulldozers and pesticides. Pandemics, embedded as they are in globe-spanning social-ecological webs, are the world's response to our shouted lectures. Do we know how to listen, to respond in creative ways, to respond again to the responses, to understand how we are changed even as we change that which is around us and defines us? What is the language that will enable us to converse in other than the most instinctual, trivial, brutal, bullying, and dysfunctional manner? Is this the as-yet-unknown language we need to tell our collective story? If this seems far-fetched as a way to think about pandemics and zoonoses, it is perhaps because we are still, as a species, so young. Nevertheless, in the midst of this very post-normal pandemic, we are learning.

One of the lessons from the 2020 COVID pandemic is that, rather than aspiring to be "lords and possessors" of nature, perhaps we should strive to be surfers, riding and adapting to the movement of forces we cannot control. But how?

In *1001 Arabian Nights*, Sultan Schahriar feels betrayed by his first wife. He seeks his revenge on women in general by having his grand vizier present him with a fresh bride every night and then having her strangled (by the grand vizier) the next morning. Scheherazade, beautiful daughter of the grand vizier, decides to put herself on the line for the sake of all women. In a successful, non-violent stratagem, she recruits her less-than-beautiful sister, Dinarzade. Every day, just before dawn, her sister awakens her and asks Scheherazade to tell her a story. The sultan is, of course, listening in and is left at dawn wanting to hear more. Every day, the sultan decides to spare her, and eventually he falls in love, first with Scheherazade's stories and then with her; he abandons his brutal, tyrannical, obsessive plans. She lives a long and meaningful life and is celebrated by young people and peace activists the world over.

The tale of Scheherazade is, finally, a tale about all of us. The earth, like Schahriar, has cut the heads off many species before us. Global history is replete with sudden or slow mass extinctions. The earth is literally built from the bones and decomposed molecules of our forebears. We humans, too, have betrayed our hosts and the bacteria who collaborated to make us possible. If we are not soon to go the way of the glyptodont and the pterosaur, then our global human family needs a good dose of psychotherapy—not just any therapy, but a narrative therapy, in which we have reimagined and retold our tale as one of survival, justice, ecological at-home-ness, and conviviality. To find this story requires a (re)search effort beyond anything we have tried before; like Scheherazade, our lives depend on it.

All of us, as global citizens, are participants in this narrative. We are all Scheherazade, the beautiful storyteller, and Dinarzade, the one who elicits this insurrection of marginalized stories from all over the world. These stories are not just of people but also of every living thing and of the earth itself. The stories in this book have been about diseases, diseases that, like Pulcher's chickens, are a voice, a reminder, a calling. The agents that cause disease in people and other animals and plants are thriving communities of tiny living beings; diseases are the consoling and horrific reminder that we are not alone, that our story has many voices, and that sometimes, if we ignore them, the tiniest of voices will inhabit our bodies and speak to us of mysteries we have ignored.

Claudius Pulcher no doubt lost his battle at Drepana in 249 BC as the result of the usual mix of bad timing and poor maritime skills. Still, one wonders about the sacred chickens that refused to eat. We don't have to believe that birds are omens, as part of some complicated spiritual mathematics, for us to take home important messages. If I were to articulate what I think Pulcher's Omen is, it is this: that all natural events tell us something well beyond the immediate causes we identify, and that if we observe carefully, we might yet survive another millennium here.

As I observed in my own work in the Caribbean and Central America, East Africa, Asia, and Canada, resolution of complex problems requires good theory, risky social engagement, and a willingness to live creatively with unresolved arguments and tensions. In these community-based ecosystem approaches to health, drawing on the best evidence from conventional science, even as we struggle with the experiences and power dynamics in communities, there has always been tension, and there are never any final, definitive experiments. From a veterinary point of view, a dead animal has all the same chemical and structural elements as a living one. The main difference between roadkill and a live dog is that in the live one, there are constant, dynamic, unresolved tensions.

We may have more sophisticated, scientific understanding of the world than ever before, but where are the stories—and more than that, where is the collective, global narrative—that can help us make sense, and give meaning to the storms of facts and information and opinions that assail us? And if we cannot agree on a grand global narrative, what, at the very least, are we looking for in this novel of novels? On what resources can we draw? Social solidarity? Love? Experience? Poetry? Nietzsche famously proclaimed that "Without music, life would be a mistake" and also that "We should consider every day lost on which we have not danced at least once." All good, but where is the story in this? Perhaps the story, a collective story encompassing all of these things, is yet to be told; we are telling it with how we live, and how we listen to each other.

Albert Camus once declared that there is more to admire in people than to despise. In the midst of the COVID-19 pandemic, after decades of snarling, misanthropic uprisings of populist nativism and many religious leaders reviving the worst strands of their complex traditions, I would rephrase that. There is at least as much, and probably more, to admire in people as there is to despise.

The global lockdowns related to COVID-19 would seem to bear this out. With the availability of Internet resources, scientists, academic institutions, private companies, and governments have openly shared information in ways that were unimaginable even five years ago. And although there have been racist attacks against Asian-looking people, and continued slurs on immigrants, legal or otherwise, many more stories have emerged of people singing together from balconies in Italy, or on YouTube, sharing home-education tips, and delivering groceries to people under quarantine. Undocumented immigrants have continued to gather our food crops, and newly arrived immigrant health workers have put their lives on the line for others. If some of the solidarity and insights provoked by COVID-19 can be carried forward into the narratives that shape the post-pandemic world, we will all be better off.

Homo sapiens, the Latin name given to our species by self-important European aristocrats, may have been wishful thinking. "Wise man" is certainly not a name based on any scientific evidence of our wisdom. Perhaps we can reimagine wisdom as a goal to which our species can aspire. We are a gift from nature that she will take back to herself when we die and the microbes recycle us. Bringing together our best political, religious, ecological, evolutionary, and philosophical stories with the artifacts we unearth through our best scholarly investigations, our work is to jostle and sing and shout and whisper them into something greater than the sum of the data, a multilayered, multiperson, multiethnic, multispecies tale of life on this planet.

The natural history of zoonoses tells us that our struggle with infectious diseases is not a war—or, if it is, it is a war against ourselves. The microbes are all around us, in us, and in our animal companions on this planet. Our struggle is ultimately one of global solidarity and keen, careful ecological awareness. I was once informed, by someone concerned about overpopulation and the limits to growth, that, as an epidemiologist, by saving people's lives, I was part of the problem. In the end, there is no logical or scientific study that proves we should care about ourselves or about the planet we live in and the other living beings with whom we share it. The care we take is a moral stance.

The ancient Indo-European word for "earth," *dhghem*, gave us not only the word "human" but also the word "humus," the organic component of soil created by bacteria, as well as the words "humble" and "humane." In this new task, we are all observing the world, collectively, as if for the first time, and constructing meaning from it. In the age of gun-toting peace-makers, God-fearing hell-raisers, and utopian pillagers, at a time when all the previous global stories have brought us rack and ruin, this is no small task. The good news is that, community by community, the world over, in the midst of a fearful pandemic, a novel global vision is emerging, revealing a planet full of complexity and mystery and beauty we never suspected was there, a world unexplored because we lacked the theoretical and practical tools. From this novel vision is emerging a global novel, full of stories, intriguing and wonderful tales hiding in the shadows of our grand delusions. The biosphere might yet spare us. Or, if the stories don't save us, they will at least, on our deathbed, in the last days of our species, have us saying, "We did quite all right, didn't we? We left a story of ourselves worth telling the universe."

SELECTED READINGS

Acha, P.N., and B. Szyfres. 2003. *Zoonoses and Communicable Diseases Common to Man and Animals*. 2nd ed. Washington, DC: Pan American Health Organization.

Brock, Arthur J., ed. and ann. 1972. *Greek Medicine: Being Extracts Illustrative of Medical Writers from Hippocrates to Galen*. New York: AMS Press.

Brothwell, D., and A.T. Sandison, eds. 1967. *Diseases in Antiquity: A Survey of the Diseases, Injuries and Surgery in Early Populations*. Springfield, IL: Charles C. Thomas.

Brown, V.A., J.A. Harris, and D. Waltner-Toews. 2019. *Independent Thinking in an Uncertain World: A Mind of One's Own*. London: Earthscan/Routledge.

Burnet, Sir M., and D.O. White. 1972. *Natural History of Infectious Disease*. Cambridge: Cambridge University Press.

Charron, D., D. Waltner-Toews, A. Maarouf, and M.A. Stalker. 2003. "A Synopsis of Known and Potential Diseases Associated with Climate Change," Ontario Forest Research Information Paper no. 154. S. Greitenhagen, and T.L. Noland (comps.). Sault Ste. Marie, ON: Ontario Forest Research Institute, Ontario Ministry of Natural Resources.

Clutton-Brock, Juliet. 1981. *Domesticated Animals from Early Times.* London: Heinemann, British Museum (Natural History).

———. 1999. *A Natural History of Domesticated Mammals.* 2nd ed. Cambridge: Cambridge University Press.

Committee on the Applications of Ecological Theory to Environmental Problems, Commission on Life Sciences, National Research Council. 1986. *Ecological Knowledge and Environmental Problem-Solving: Concepts and Case Studies.* Washington, DC: National Academy Press.

Corvalan, C., S. Hales, and A.J. McMichael, eds. 2005. *Ecosystems and Human Well-Being Synthesis.* Geneva: WHO.

Desowitz, Robert S. 1981. *New Guinea Tapeworms and Jewish Grandmothers: Tales of Parasites and People.* New York: Avon Books.

Evans, R.G., M.L. Barer, and T.R. Marmor. 1994. *Why Are Some People Healthy and Others Not?* New York: Aldine de Gruyter.

Fiennes, Richard N. 1978. *Zoonoses and the Origins and Ecology of Human Disease.* London: Academic Press.

Garrett, Laurie. 1994. *The Coming Plague.* New York: Penguin Books.

Hyams, Edward. 1972. *Animals in the Service of Man: 10,000 Years of Domestication.* London: J.M. Dent & Sons.

Institute of Medicine. 1992. *Emerging Infections: Microbial Threats to Health in the United States.* Washington, DC: National Academy Press.

Krauss, Hartmut, Albert Weber, Max Appel, Burkhard Enders, Henry D. Isenberg, Hans Gerd Schiefer, Werner Slenczka, Alexander von Graevenitz, and Horst Zahner. 2003. *Zoonoses: Infectious Diseases Transmissible from Animals to Humans.* 3rd ed. Washington, DC: ASM Press.

Lefebvre, S.L., R.J. Reid-Smith, D. Waltner-Toews, and J.S. Weese. 2009. "Incidence of Acquisition of Methicillin-Resistant *Staphylococcus aureus, Clostridium difficile*, and Other Health-Care-Associated Pathogens by Dogs that Participate in Animal-Assisted Interventions." *J Am Vet Med Assoc.* 234:1404–17.

Margulis, Lynn, and Dorion Sagan. 1986. *Microcosmos: Four Billion Years of Evolution from Our Microbial Ancestors*. New York: Summit Books.

Marrie, Thomas, ed. 1990. *Q Fever, Volume 1: The Disease*. Boca Raton: CRC Press.

McMichael, A.J. 1993. *Planetary Overload: Global Environmental Change and the Health of the Human Species*. Cambridge: Cambridge University Press.

McNeill, W.H. 1976. *Plagues and Peoples*. Garden City, NY: Anchor Books.

Palmer, S.R., Lord Soulsby, and D.I.H. Simpson. 1998. *Zoonoses: Biology, Clinical Practice and Public Health Control*. Oxford: Oxford University Press.

Pavlovsky, E.N. 1966. *Natural Nidality of Transmissible Diseases; With Special Reference to the Landscape Ecology of Zooanthroponoses*. English trans. F.K. Plous Jr. Urbana, IL: University of Illinois Press.

Scheld, W.M., William Craig, and James Hughes, eds. 1998. *Emerging Infections*, vols. 1 & 2. Washington, DC: ASM Press.

Schwabe, C.W. 1984. *Veterinary Medicine and Human Health*. Baltimore: Williams & Wilkins.

Scullard, Howard H. 1960. *A History of the Roman World from 753 to 146 BC*. 3rd ed. London: Methuen.

Tarán, L., ed. *Aristotle, On the Parts of Animals*. Greek and Roman Philosophy, vol. 26. New York: Garland Publishing.

Waltner-Toews, David. 2004. *Ecosystem Sustainability and Health: A Practical Approach*. Cambridge: Cambridge University Press.

———. 2008. *Food, Sex and Salmonella*. 2nd ed. Vancouver: Greystone Books.

———. 2010. "One Health for One World." Veterinarians without Borders/ Vétérinaires sans Frontières—Canada. vsf-international.org/project/ one-health-for-one-world-a-compendium-of-case-studies-by-vsf-canada/.

———. 2013. *The Origin of Feces: What Excrement Tells Us about Evolution, Ecology, and a Sustainable Society.* Toronto, ON: ECW Press.

———. 2017. "Zoonoses, One Health and Complexity: Wicked Problems and Constructive Conflict." *Phil. Trans. R. Soc. B* 372: 20160171. doi.org/10.1098/rstb.2016.0171.

Waltner-Toews, D., and A. Ellis. 1992. *Good for Your Animals, Good for You.* Guelph, ON: Pet Trust.

Waltner-Toews, D., J. Kay, and N.-M. Lister, eds. 2008. *The Ecosystem Approach: Complexity, Uncertainty, and Managing for Sustainability.* New York: Columbia University Press.

Waltner-Toews, M., and D. Waltner-Toews. 2017. "Designed for Disease? An Ecosystem Approach to Emerging Infectious Diseases." *LA+ Interdisciplinary Journal of Landscape Architecture*, 6:88–93.

Zinsser, Hans. 1971 (8th printing of Bantam edition). *Rats, Lice and History.* Boston: Little, Brown.

Zinsstag, J., E. Schelling, D. Waltner-Toews, M. Whittaker, and M. Tanner, eds. 2015. *One Health: The Theory and Practice of Integrated Health Approaches.* Wallingford, UK: CAB International.

OTHER SOURCES AND RESOURCES

The history, characterization, and re-emergence of zoonoses are addressed in a rapidly expanding variety of scientific journals and technical reports, many of them digitized, and I have searched through most of them over the past decades for material for teaching and research. Organizations working on these issues, each with their own journals, conferences, and online resources, have expanded exponentially since the first edition of this book came out, and I will only describe those few with which I have had personal experience.

The international Resilience Alliance, brought together by ecologist C.S. Holling in the 1990s, has contributed a great deal to our ability to develop appropriate response and management strategies that will promote both ecological resilience and human well-being. There are other, similar networks, many of which have their own perspectives, purposes, and institutional supports. Various One Health platforms and initiatives, integrating animal and human health at a policy level, tend to work "top-down" and are associated with academia, as well as official national and international organizations such as WHO and its sister, the World Organisation for Animal Health (OIE). Our more loosely organized Post-Normal group focuses more generally on the philosophical, theoretical, and practical challenges of science being used for policy. The International Association for Ecology and Health (IAEH), EcoHealth Alliance, and related organizations are oriented toward research into EIDs. EcoHealth International (which grew out of, and separated from, the IAEH), is a complement to both the research and One Health initiatives, being rooted in locally driven, community-based research and activism. The Communities of Practice in Ecosystem Approaches to Health, such as COPEH-Canada and COPEH-LAC in Latin American and the Caribbean, have developed courses and training materials and linked communities, researchers, and teachers both online and in person. Veterinarians without Borders/Vétérinaires sans Frontières–Canada and VSF International/Vétérinaires sans Frontières work in communities throughout the world, integrating the health and well-being of people, other animals, and the ecosystems that support us. Having been active in most of these initiatives, I've found that they all have roles to play in creating a more sustainable, healthy planet.

These networks, toggling between on-the-ground work and theory, personal contact and social networking, are developing a new kind of engaged, active science capable of asking, and perhaps answering, the questions too big to be contained in the laboratory. Among those questions, one is central: How can a convivial human existence on the earth be created and prolonged?

ACKNOWLEDGMENTS FOR THE SECOND EDITION

IT IS a cliché, but so many people have contributed to my knowledge and understanding of zoonoses that it would be impossible to list them all, so I won't try. For a partial list, have a look at the first edition of this book. For the second edition itself, thanks to Rob Sanders; publishers, in my experience, rarely ask a writer if they have a book they could publish. Very special thanks to Jennifer Croll and Paula Ayer at Greystone for making this happen at the breathtaking speed of the SARS-COV-2 pandemic itself. And thanks to Greystone's art director, Nayeli Jimenez, and proofreader Alison Strobel for making sure the book was washed, coiffed, dressed up, and ready to go out on a pandemic spring date.

Parts of the section on West Nile virus and landscape design in chapter 8 were first published as: Matthew Waltner-Toews and David Waltner-Toews, "Designed for Disease? An Ecosystem Approach to Infectious Diseases," *LA+ Interdisciplinary Journal of Landscape Architecture* no. 6 (University of Pennsylvania, 2017): 88–93.

INDEX